Open Autoclave: Build an open-source off-grid medical instrument sterilizer

by David Hartkop

Written 2018 by David Hartkop, Idea Propulsion Systems

As the author, I choose to release the entire contents of this publication into the creative commons through an Attribution CC-BY license.

You are free to:

Share, copy and redistribute the material in any medium or format.

Adapt, remix, transform, and build upon the material for any purpose, even commercially. The licensor cannot revoke these freedoms as long as you follow the license terms:

Under the following terms:

Attribution: You must give appropriate credit, provide a link to the license, and indicate if changes were made. You may do so in any reasonable manner, but not in any way that suggests the licensor endorses you or your use.

No additional restrictions: You may not apply legal terms or technological measures that legally restrict others from doing anything the license permits.

If you wish to contribute monetarily toward this or future projects by the author, you may purchase printed editions or donate through the author's website at the following:

www.ideapropulsionsystems.com/OpenAutoclave

Hartkop, David T.

Open Autoclave: Create an open-source off-grid medical instrument sterilizer

ISBN: 978-1729731949

For more information:

Idea Propulsion Systems
4070 Willow Springs Rd.
Central Point OR 97502

The Maker Movement

The Maker Movement in Brief

People who make things are makers. When makers share their creations through the internet, they create maker culture. When maker culture drives business startups, the flow of goods and services, and the philosophy of human endeavor, the culture becomes a movement.

The maker movement is a fertile valley of possibilities that draws from the fields of design, electronics, computer science, materials, performance, communication, and art. It is a playground for new ideas in business, work, art, and play. The creation of shared open-source hardware, software, and instructions has produced a body of capability that is accessible to many billions of people around the world.

Online resources such as Pinterest, YouTube, Facebook, Make Magazine, Instructables, Tinkercad, Thingiverse, Adafruit, Sparkfun, Github, Blender, Arduino and Python are all parts of this movement. They offer tools for creating as well as places to collaborate. The maker movement, however, is more than the sum of these resources. It is a living thriving community of people who create and share. Each time a maker shares his or her project, the community's overall capability grows - the fertile valley becomes richer.

It should be said that the movement is not about working for free - there is money to be made, there are businesses to start, and technologies to roll out. The philosophy behind maker culture, however, says that there is value in sharing ideas because it improves the overall ecosystem of innovation. When tools are shared and improved together, everyone has access to better tools than they can afford independently. A might state his or her philosophy this way: Do your own thing, but also share knowledge that improves the world for other makers.

The Rise of Humanitarian Makers

While the maker movement has roots in entertainment and hobby-crafts, makers are increasingly looking for ways to apply their ideas and skills to solving real-world problems. Shared knowledge of design and technology are being used to save lives, to alleviate suffering, and to restore human dignity. These humanitarian aspirations take many forms because they address problems in fields as diverse as education, healthcare, disaster relief, communications, energy production, water availability, transportation, commerce and more.

Examples of humanitarian maker projects include low cost water filters, phone apps for coordinating rescue efforts, open-source medical equipment, and LED reading lamps powered by gravity.

This book is a Humanitarian Maker Project

This book teaches readers how to build a low-cost medical autoclave for sterilizing medical instruments. It is designed to run off-grid and away from conventional medical resources. It plugs into a motor vehicle's 12-volt cigarette lighter. It can also be solar, or wind powered. It can be made by healthcare workers and then deployed to rural medical clinics and mobile treatment centers where needed. By distributing these practical, cost-effective systems around the world, the burden of disease due to infection can be reduced, saving many lives in the process. That is the goal.

Making this goal a reality: By building and testing this system, you are independently confirming and correcting the design and the instructions in this book. If you have improvements, share them! Your participation will improve the design, which translates into better performance, higher ease of construction, lower cost, and greater awareness of the technology in the healthcare community. Ultimately all these improvements translate into more lives saved.

Table of Contents

Introduction 2

Chapter 1 - Setup
1.1 Set up your workspace 5
1.2 Tools 5
1.3 Parts and supplies 6

Chapter 2 - Build the autoclave oven
2.1 Insulated pot 16
2.2 Base tray 19
2.3 Heater rack 21

Chapter 3 - Build the electronic controller
3.1 Drill holes in the electronics box 25
3.2 Install the components 30
3.3 Finish the wiring 39

Chapter 4 - Set up the controller computer
4.1 Install the Arduino software 47
4.2 Download the Open Autoclave software.............49
4.3 Upload the software to the microcontroller 51

Chapter 5 - Test the autoclave system
5.1 Run a full cycle 55
5.2 Download and graph the data 59
5.3 Using biological indicator strips 62
5.4 Troubleshooting..63
5.5 Using the autoclave, no computer 65

Chapter 6 - Resources
6.1 Change the temp and time settings 68
6.2 Volt-ohm meter 69
6.3 Anti-static mat 70
6.4 Pin connector crimper 71
6.5 Full electrical schematic..72
6.6 Diagram of autoclave system 73
6.7 Template for drilling electrical box lid 74
6.8 Template for drilling electrical box 75
6.9 Placement of components in electrical box.........76
6.10 Ideas for improvements 77
6.11 Powering Open Autoclave with Solar................78

Acknowledgements 79

Share Your Experience 80

Introduction

0.1 About the project

0.1.1 The need for autoclaves

According to a World Health Organization (WHO) study, people having surgery in developing nations are up to nine times more likely to contract infections during surgery than people in the developed world. Pregnancy complications and other surgical diseases account for 11% of the Global Disease Burden.

Many of these infections can be avoided by using proper sterilization techniques. One of the most basic and successful sterilization technologies in use today is the medical autoclave - essentially an oven for "cooking" medical instruments before use. By exposing medical instruments to high temperatures, bacteria and bacterial spores can be eliminated from medical instruments. Autoclaves, however, require a stable electrical energy supply and are expensive to purchase and operate.

Worldwide, over 3 billion people in developing countries are served by rural clinics, but more than half of these clinics lack access to electricity. Even in areas with access to energy, a typical medical sterilizer costs between $1200 USD and $5000 USD, a cost often too great to shoulder.

Inability to properly sterilize medical instruments puts many common medical procedures entirely out of reach - the high risk of infection is too great. Every year there are 50-60 million people in developing countries who suffer from wound injuries that would benefit from surgery. Considering a small subset of these cases, WHO estimates that the lives of 500,000 women could be saved annually with basic surgical interventions during childbirth.

Many thousands of lives can be saved and improved by providing the means to sterilize medical instruments to rural health clinics. The Open Autoclave offers a practical off-grid solution at a fraction of the cost of systems with similar capability. It is based on the tools, materials and knowledge base common to maker culture, and instructions for making it are free and open-source.

Let's work together to make medical autoclaves available to those who need them the most.

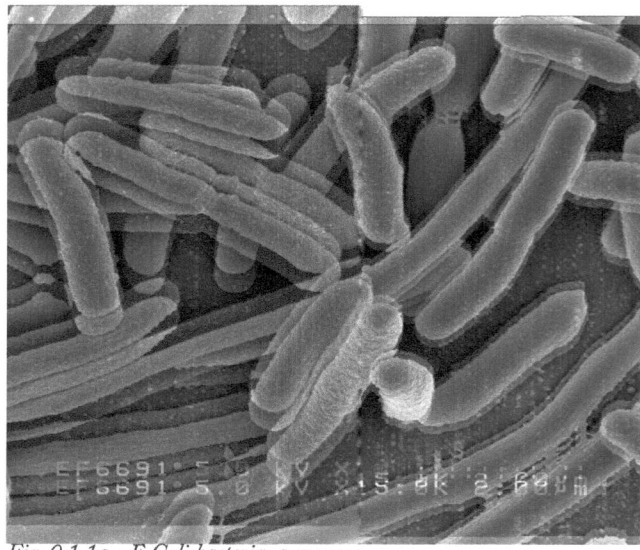

Fig. 0.1.1a - E.Coli bacteria, a common cause of infection

REFERENCES

1 Speigel, David A. MD. Health Volunteers Overseas: WHO and Essential Surgical Care. July 2010.

2 Weiser, Thomas G., et al. An estimation of the global volume of surgery: a modeling strategy based on available data. The Lancet: Vol 372 July 12; 2008, 139-144.

3 Centers for Disease Control, Guidelines for Disinfection and Sterilization in Healthcare Facilities 2008.

4 WHO, Emergency and Essential Surgery: the backbone of primary health care, http://www.who.int/eht/sb/en/ (accessed on 26 Aug 2010)

5 Komp, R., et al. Solar autoclaves for sterilizing medical instruments in remote settings.

6 Knowlton, Lisa M., A geospatial evaluation of timely access to surgical care in seven countries, Bulletin of the World Health Organization, Volume 95, Number 6, June 2017, 389-480

7 Allegranzi, Benedetta MD, Burden of endemic health-care-associated infection in developing countries: systematic review and meta-analysis, The Lancet VOLUME 377, ISSUE 9761, P228-241, JANUARY 15, 2011 Published:December 10, 2010DOI:https://doi.org/10.1016/S0140-6736(10)61458-4

8. D. Tao, Gregory & S. Cho, Hallie & Frey, Daniel & G. Winter, Amos. (2012). Design of a Low-Cost Autoclave for Developing World Health Clinics. Proceedings of the ASME Design Engineering Technical Conference. 7. 10.1115/DETC2012-71435

9. Pittinger, Matt, Distributed surgical instrument sterilization using SOLAR POWERED AUTOCLAVES in low resource settings (2010) WHO, Innovative technologies that address global health concerns: outcome of the call global initiative on health technologies, 2010, Section 3.5.14

10. Medical autoclave exposure time figures published by JADA, (Journal of American Dental Association) Vol. 122 December 1991:

Introduction

Fig. 0.1.1c - Finished OpenAutoclave, ready for use.

0.1.2 Description of this project

This publication describes a 12-volt dry-heat autoclave that can be constructed from widely manufactured supplies for less than $200 USD.

The autoclave can be powered directly from an automotive cigarette lighter port, or via alternative energy sources such as solar or wind power.

An Arduino microcontroller is used to regulate autoclave temperature over time, and to record data logs, which may be downloaded to a laptop computer.

The autoclave can be built and used without a computer but will then require attention to temperature and time on the part of the user.

While writing this book, I put considerable effort into testing the build instructions. I verified the success of the autoclave using mail-in biological indicator strips purchased from a laboratory quality control service. Heating the strips in the autoclave kills the bacterial spores, proving that it will work to sterilize medical instruments.

There is no warranty or guarantee that your system will perform as described, so I encourage you to test your autoclave in a similar way. There are many providers of sterilizer monitoring services worldwide. You need a lab that can test Dry Heat process autoclaves.

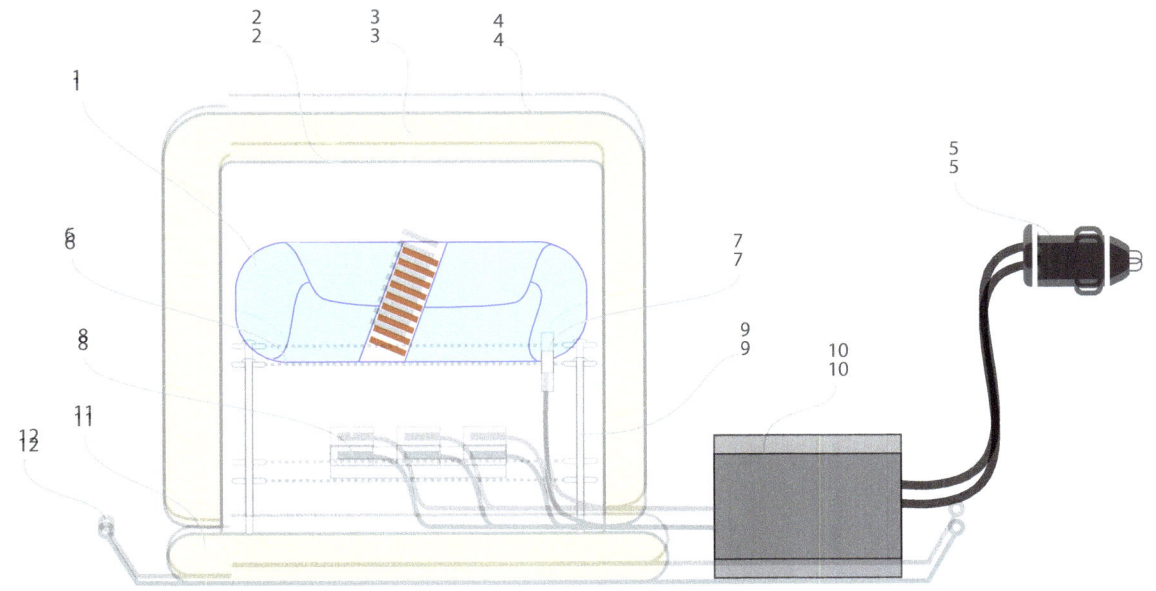

1: Wrapped medical instruments
2: Inner stainless stock pot
3: Ceramic fiber insulation
4: Heavy aluminum foil and tape outer wrap
5: 12 VDC power input connector
6: Aluminum pizza screen inner stages
7. Stainless thermocouple temperature probe
8. 40 Watt heater cartridge
9. Stainless support hardware
10. Microcontroller
11. Ceramic fiber bottom pillow
12. Full size commercial cookie sheet, aluminum

Fig. 0.1.2 - Labeled Cross Section of OpenAutoclave system

Introduction

0.1.3 YOU can build this autoclave :-)

This project is for motivated healthcare workers with technical skills. It is also accessible to engineering students and educators who are looking to apply knowledge of electronic control systems. The project can also be successfully completed by humanitarian workers or volunteers interested in creating useful equipment for relief efforts. It requires patience and attention to detail but is not overly complex to build or troubleshoot.

0.1.4 Free and open source Information

As the author, I choose to release the entire contents of this publication into the creative commons through an Attribution CC-BY license.

You are free to:

Share, copy and redistribute the material in any medium or format.

Adapt, remix, transform, and build upon the material for any purpose, even commercially.

The licensor cannot revoke these freedoms as long as you follow the license terms.

Under the following terms:

Attribution: You must give appropriate credit, provide a link to the license, and indicate if changes were made. You may do so in any reasonable manner, but not in any way that suggests the licensor endorses you or your use.

No additional restrictions: You may not apply legal terms or technological measures that legally restrict others from doing anything the license permits.

If you wish to contribute monetarily toward this or future projects by the author, you may purchase printed editions or donate through the author's website at the following:

www.ideapropulsionsystems.com/OpenAutoclave

Chapter 1 - Setup

1.1 Set up your workspace

The Open Source Autoclave is designed to be created with only hand tools and without need for a full workshop. You will, however, need a clean well-lit table area for your project. Most steps can be done with nothing more than a screwdriver. Some other steps require a cordless drill, pliers, and a small utility knife.

Be sure to use judgment when allowing children access to tools. The only part of the process that may require some outdoor work is the necessity to apply ceramic fiber insulation wrap. Ceramic fiber sheds fine powder, and so it should only be handled by adults wearing dust masks and appropriate eye protection. It is easier to clean up if done outdoors.

Safety Wear:

    Dust mask

    Eye protection

    Plastic gloves

    Hearing Protection (for grinder if used)

1.2 Tools

Cordless electric drill

Drill bits:

    1/8 (3 mm)

    17/64 (7 mm)

    1/2 (13 mm)

Sharp punch for marking for drilling

Needle nose pliers

Hacksaw or electric cutting wheel

Hand pliers

Ruler (inches & cm)

Tape measure, inches or metric

Scalpel or utility knife

Scissors

Screwdrivers (Philips & regular, large & small)

Set of hex keys, or Allen wrenches (metric)

Volt-ohm meter for double checking

Wire cutters

Wire strippers

Soldering iron & solder roll (optional)

Wire connector crimping tool (optional)

Fig. 1.2 - Assorted hand tools used to build the OpenAutoclave

Chapter 1 - Setup

Section 1.3 Supplies

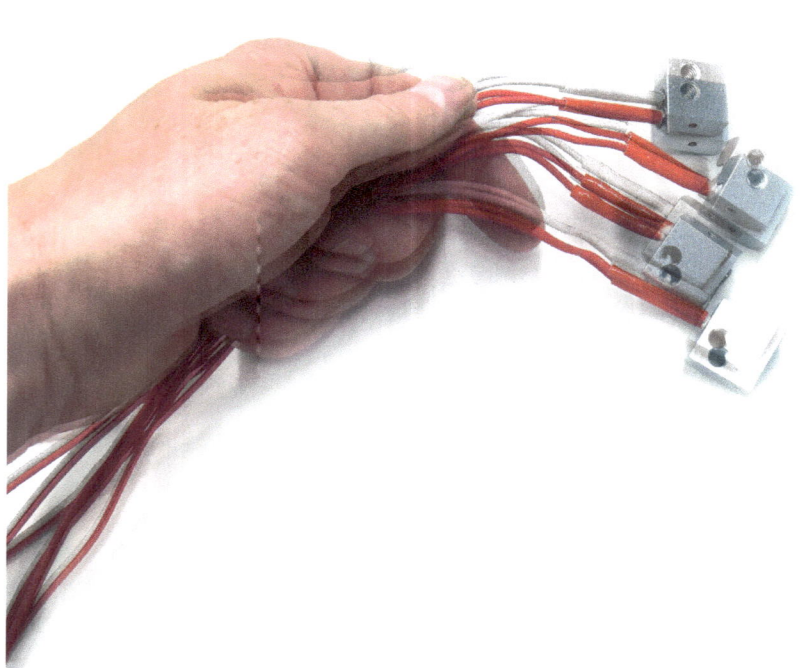

Fig. 1:3 - 12v heater cartriges in aluminum mounting blocks

Collecting all the parts is the most time-consuming part of this project. The supplies break down into these three categories:

- Oven Parts
- Heater Rack Parts
- Controller Parts

Each and every part is detailed on the following pages. Many of the parts are so common that there are several different manufacturers.

Some of the parts can be purchased in retail home improvement or kitchen supply stores. Everything can also be purchased online via big online retailers like Amazon, Walmart, AliExpress, or even eBay.

Chapter 1 - Oven Parts

Half-Sheet Aluminum Sheet Pan

Used for: The base of the system

Quantity: 1

Details: 13" x 18" Commercial 18 Gauge Aluminum Sheet Pan

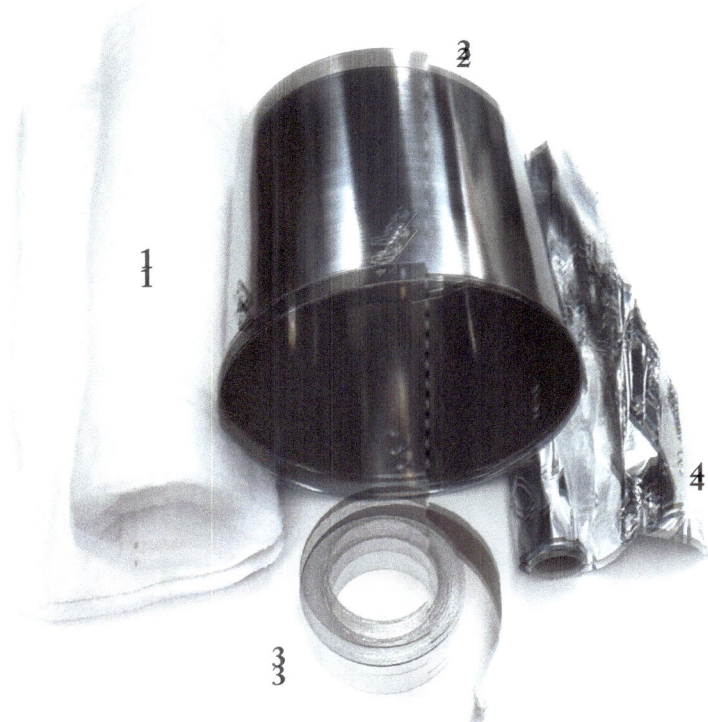

1. Ceramic Fiber Insulation

Used for: Insulating the autoclave oven

Quantity: 1 roll of material: (1 x 24 x 36 in)

This is enough to cut three shapes:
- 9.5 x 33 in (24cm x 84cm) rectangle
- 12 in (31cm) circle
- 18 x 18 in (46cm x 46cm) square

Details: One roll of Ceramic Fiber Blanket 8# Density, 2300°F

2. Stainless Stock Pot (with handles cut off)

Used for: The body of the autoclave oven

Quantity: 1

Details: 12-quart stainless stock pot (Mfg. #:MS14-042-420-45)

Size: (LxWxH): 10.04 x 10.04 x 9.45 Inches

3. Heavy Aluminum Foil Roll

Used for: Wrapping and protecting the insulation

Quantity: 1

Details: Reynolds Wrap Heavy-Duty Aluminum Foil (75 sq. ft)

4. Aluminum Furnace Tape Roll

Used for: Sealing the foil wrapped around the insulation

Quantity: 1

Details: Scotch aluminum foil furnace tape 2-inch width.

Chapter 1 - Heater Rack Parts

Cartridge type heating elements

Used for: heating elements for the autoclave

Quantity: 4

Details: Set of 6mm dia. x 20mm long elements, rated 12V 40W These are Ceramic Cartridge Heaters for 3D Printer "Reprap Prusa" (4 are required, best to purchase a pack with some extras.)

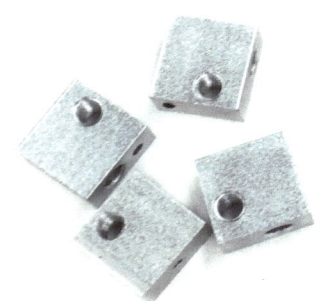

Aluminum heater blocks

Used for: mounting the heating elements

Quantity: 4

Details: Aluminum Heater Block Specialized for MK7 MK8 3D Printer Extruder (purchase as a pack with extras)

Iron pipe floor flange

Used for: heat radiator within the autoclave oven

Quantity: 1

Details: 1/2-inch black malleable iron threaded floor flange pipe fitting

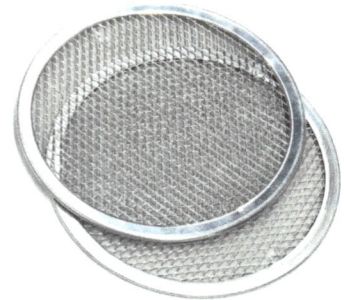

Aluminum pizza screens

Used for: Supporting the heating elements and wrapped medical instruments inside the autoclave

Quantity: 2

Details: American Metalcraft 8 in (20 cm) diameter expanded metal (seamless) aluminum pizza screen or similar

Automotive cigarette lighter cable

Used for: providing power to the system from a 12-volt source

Quantity: 1

Details: 12V Heavy Duty 16 AWG 15A Male Plug Cigarette Lighter Adapter Power Supply Cord with 1 Meter (3.3 Feet) Cable Wire or similar. System pulls 7.5 amps at 12 volts.

Chapter 1 - Heater Rack Parts

Wire nuts

Used for: easily wiring together parts of the power system

Quantity: 7

Details: Twist-on wire nuts suitable for 12-18 Gauge wire

Long bolts for heater rack

Used for: the legs of the heater rack

Quantity: 4

Details: 1/4-20 x 4 in long (M6 x 10 cm) stainless carriage bolt

Nuts & washers for heater rack

Used for: securing pizza screens to leg bolts

Quantity: 16 of each

Details: 1/4-20 (M6) stainless nuts & stainless washers. (Zinc coatings can burn, releasing toxic fumes!)

Screws

Used for: mounting heater elements to pizza screen

Quantity: 4

Details: M6 x 20 mm, stainless, tapered head.

Nuts

Used for: mounting heater elements, thermocouple to screen

Quantity: 5

Details: M6 stainless nuts

Chapter 1 - Controller Box Parts

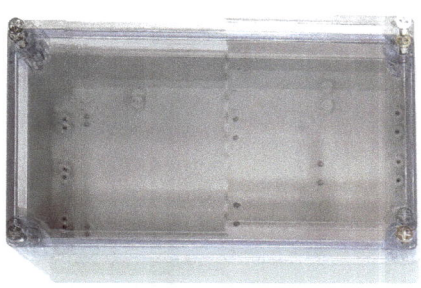

Electrical enclosure
Used for: Holding all the electronics for the control system
Quantity: 1
Details: 6.2 x 3.5 x 2.3 in ABS Waterproof Junction Box Universal Project Enclosure with Transparent PC Cover

Arduino Uno R3 microcontroller
Used for: monitoring temperature and time of autoclave exposure
Quantity: 1
Details: Arduino Uno R3 Microcontroller or similar
(Spend at LEAST $12 USD on this or it will be total junk!)

Controller mounting bracket
Used for: mounting the Arduino into the electrical enclosure
Quantity: 1
Details: Arduino Uno plastic mounting bracket, comes with official Arduino Uno, or can be ordered separately.

12-volt automotive relay
Used for: controlling the heating elements to prevent overheating
Quantity: 1
Details: RLS125 12-VCB Automotive Relay SPDT 30/40A or similar

Chapter 1 - Controller Box Parts

MOSFET module
Used for: Arduino control of the Auto Relay
Quantity: 1
Details: IRF520 MOSFET driving module or similar N-Channel FET for driving a relay with ~5V (PWM) controller pin outs.

LCD screen module
Used for: displaying temperature and alerts from the Arduino
Quantity: 1
Details: 1602 16x2 Serial HD44780 Character LCD Board Display 5V with IIC/I2C Serial Interface Adapter Module for Arduino

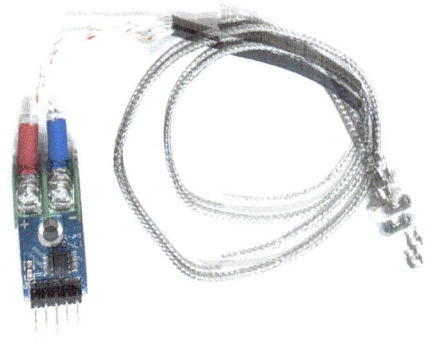

Thermocouple with amplifier module
Used for: measuring temperature inside the autoclave
Quantity: 1
Details: HiLetgo DC 3-5V MAX6675 Module + K Type Thermocouple Temperature Sensor Thermocouple Sensor Set M6 Screw for Arduino
(Other types can be used but will require changes to Arduino code.)

Heater power switch
Used for: power switch to heaters elements
Quantity: 1
Details: Heavy-Duty Electrical Toggle Switch, SPST, ON-OFF, 20 A/125V AC, Spade Terminal or similar

Chapter 1 - Controller Box Parts

Micro speaker

Used for: alert sounds to signal autoclave status

Quantity: 1

Details: Motherboard Computer PC Internal Speaker Buzzer Computer Case Buzzer for Arduino (Mfg. #: CYT1090) or similar

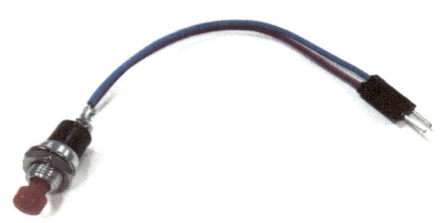

Momentary push button

Used for: restarting the autoclave-cycle timer

Quantity: 1

Details: 12mm SPST NO Reset Switch Push Button Switch Model EK1922 or similar

Hook up pin connector wires

Used for: Connecting various components to the Arduino controller

Quantity: 13

Details: M to F (M/F) Breadboard jumper wires AKA Dupont pin connector jumper wires, 10 cm length. Total needed: 13, see details below:

x4 for LCD

x2 for Speaker

x2 for Push Button

x3 for MOSFET module

x2 for Power Supply

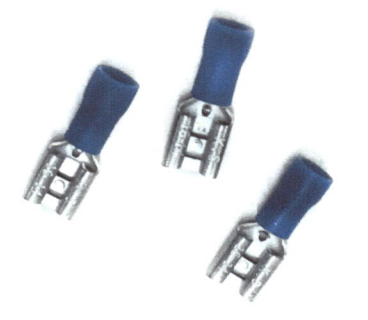

Female spade crimp connectors

Used for: Connecting wires to the automotive relay and also to the heater power switch.

Quantity: 6

Details: For automotive relay hookup, power switch; Female Spade Crimp Wire cable connector blue (6.3 mm) for 14-16 gauge wire

Chapter 1 - Controller Box Parts

Short screws

Used for: mounting Arduino to Bracket, MOSFET to Enclosure

Quantity: 5

Details: 4-40 x 0.25 in (M3 x 6 mm) zinc coated machine screw

Medium screws

Used for: mounting Arduino bracket, through enclosure, to aluminum tray

Auto Relay to Enclosure

Quantity: 5

Details: 4-40 x 3/8 in (M3 x 10mm) zinc coated machine screw

Long screws

Used for: LCD to box lid

Quantity: 4

Details: 4-40 x 0.75 in (M3 x 20 mm) zinc coated machine screw

Nuts

Used for: securing LCD to box lid, Auto relay to enclosure, Arduino bracket through enclosure to aluminum tray

Quantity: 13

Details: 4-40 (M3) zinc coated nuts

Open Autoclave: Build an open-source off-grid medical instrument sterilizer by David Hartkop CC-BY license

Chapter 1 - Setup

Tiny washers
Used for: mounting auto relay to enclosure, Arduino bracket through enclosure to aluminum tray
Quantity: 5
Details: 4-40 (M3) zinc coated washers

Speaker Wire Pair
Used for: connecting controller to heater elements
Quantity: 12 in (30 cm)
Details: 16 gauge, two-conductor stranded copper wire

USB printer cable
Used for: programming the Arduino, Downloading data.
Quantity: 1
Details: 36-inch USB 2.0 cable male-male A to B

Chapter 2 - Build the Autoclave Oven

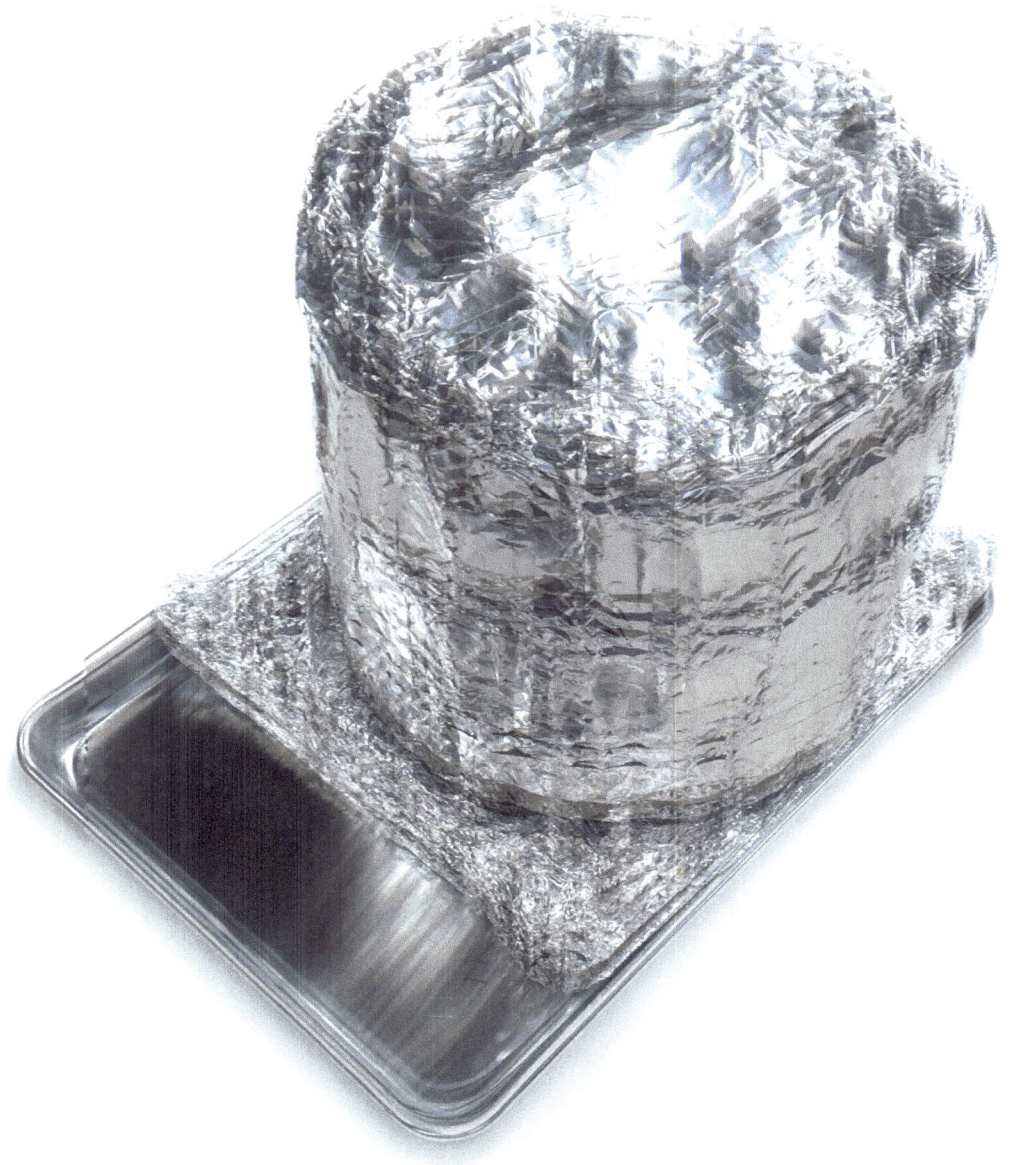

Fig. 2.0 – Insulated oven made from a stock pot, a cookie sheet, ceramic fiber insulation, foil, and foil tape

2.1 **Insulated pot**
2.2 **Base tray**
2.3 **Heater rack**

Chapter 2 : Section 1 - Insulated pot

2.1.1 Cut the handles off a stainless steel pot

Cut the handles off a stainless pot. Use a hack saw or an electric handheld grinder with a cutting wheel. Trim the handles as flat as possible but don't remove the rivets.

Pictured at left is the pot with the handles removed, along with the other supplies for this section.

2.1.2 Cut some pieces of ceramic fiber insulation

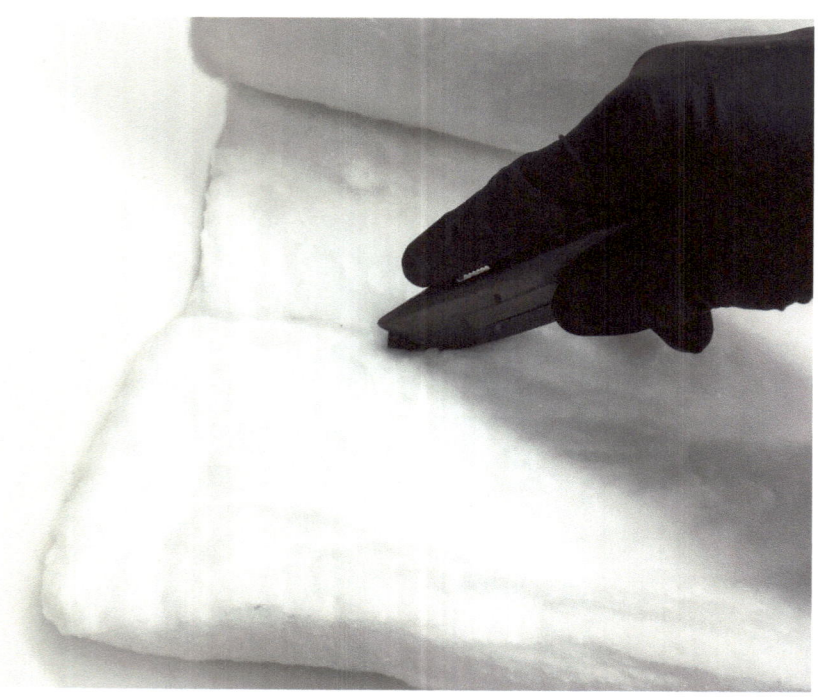

The following step assumes your pot size is (LxWxH): 10.04 x 10.04 x 9.45 Inches. Adjust accordingly if you have a different sized pot!

Use a tape measure and utility knife to cut three pieces of insulation:

Cut a 9.5-inch x 33-inch (24cm x 84cm) rectangle to wrap around the pot.

Cut a 12-inch (31cm) diameter circle to cover the round end of the pot.

Cut an 18 x 18-inch (46 cm x 46 cm) square to be used in step 1.2.1 for the tray's 'pillow.'

Important: Wear a dust mask and gloves for this step.

Chapter 2 : Section 1 - Insulated pot

2.1.3 Insulate the pot: Foil, then insulation, then foil

a. Wrap the bare pot with heavy foil. Wrap it in a way that leaves an open skirt of foil that stands out from the pot by 10 cm (4 inches) or so, as shown in the picture 1.1.3a. This skirt lets the inner foil wrap be easily seamed together with the outer wrap, fully containing the insulation.

b. Next wrap with the insulation. First wrap the cylindrical outside of the pot with the 9.5-inch x 33-inch rectangular section of insulation. Temporarily secure it with some furnace tape around the outside.

Next, place the circular 12-inch diameter piece of insulation onto the top to make a cap.

c. Now, wrap heavy foil around the entire outside. Foil can be seamed together by rolling over lapping edges tightly. This contains any potential shedding from the insulation.

2.1.4 Wrap the entire pot with heavy foil and furnace tape

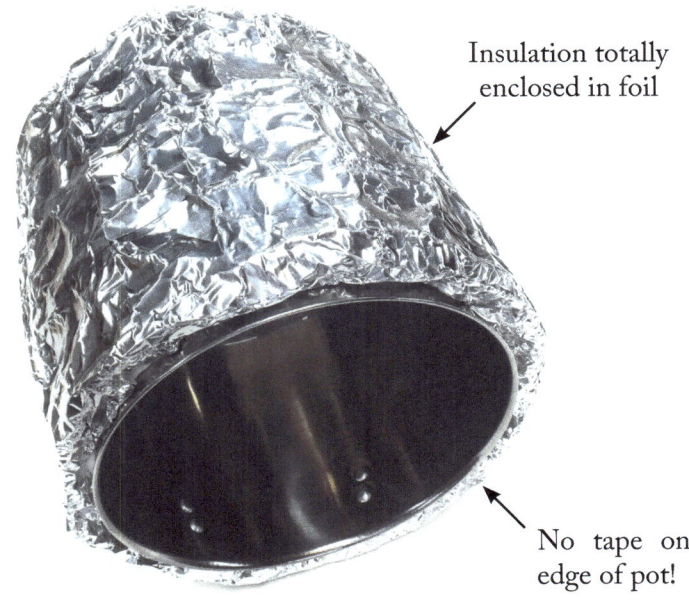

Finally, wrap the entire pot with a final layer of heavy foil. Cover all seams with aluminum furnace tape. The finished insulated pot should be fully wrapped with no insulation showing anywhere, as shown in figure 1.1.4.

Note: NO TAPE INSIDE THE AUTOCLAVE CHAMBER!

Do not use any foil tape around the bottom edge. If any tape is used inside the chamber, the adhesive will burn, contaminating your tools!

Chapter 2 : Section 1 - Insulated pot

1:1.5 Install a barbeque oven thermometer for manual operation

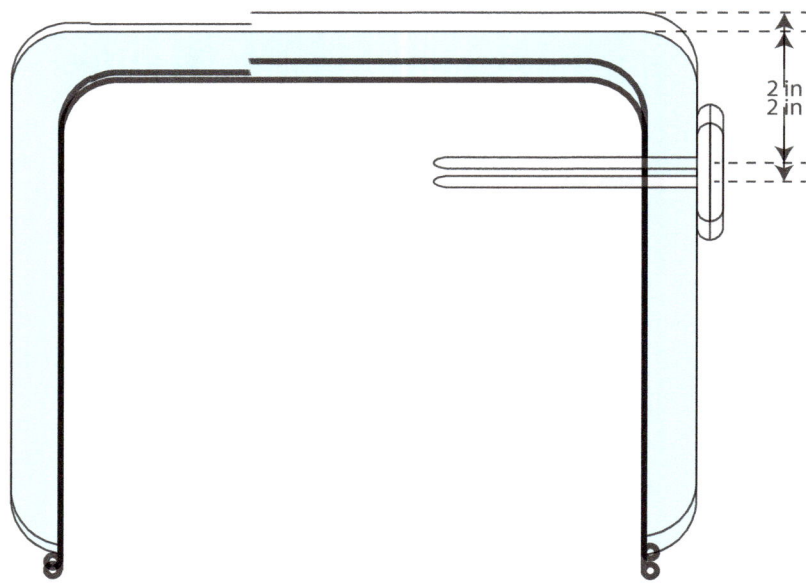

If you plan to operate your autoclave with no computer system, you will need to install an analog thermometer of the type normally used for barbeque ovens.

Select a drill bit that allows for the tightest fit possible for the probe of your thermometer.

Drill a hole into the pot approximately 2 inches (5 cm) down from the insulated end. Once the hole is drilled, insert the thermometer and secure on the outside with tape.

Figure at left shows a cross section of the arrangement.

For instructions on how to use the autoclave manually, see chapter 4 section 5 (pg. 65).

Chapter 2 : Section 2 - Base tray

1.2.1 Cut a square of insulation to make the pillow

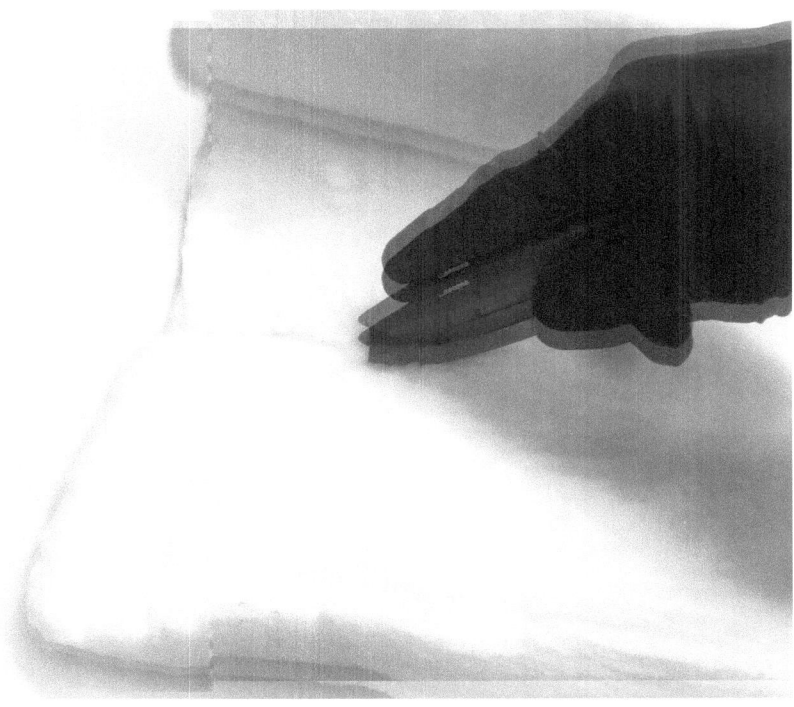

1.2.1 Cut a square of insulation to make the pillow

If you didn't already do this back in step 1.1.2, use a tape measure and a utility knife to cut an 18-inch square of ceramic fiber insulation.

Important: Wear a dust mask and gloves for this step.

2.2.2 Wrap the square in foil

Wrap this square entirely in aluminum foil, being sure to carefully roll the edges to fully contain all dust.

Chapter 2 : Section 2 - Base tray

2.2.3 Tape pillow to tray

Place the insulated pillow onto the aluminum tray as shown. Secure the edges of the pillow to the tray with aluminum tape. Finally, place the insulated pot onto the foil and gently press down to form a 'gravity seal' against the pillow.

You may wish to place a set of four self-adhesive rubber feet under each of the four corners of the aluminum tray.

NOTE: Keep all tape outside the footprint of the chamber. If any tape is inside the heated chamber, the adhesive will burn, contaminating the contents.

This completes the oven section of your autoclave.

This completes the oven section of your autoclave.

Chapter 2 : Section 3 - Base tray

2.3.1 Put 4 heater elements into heater blocks

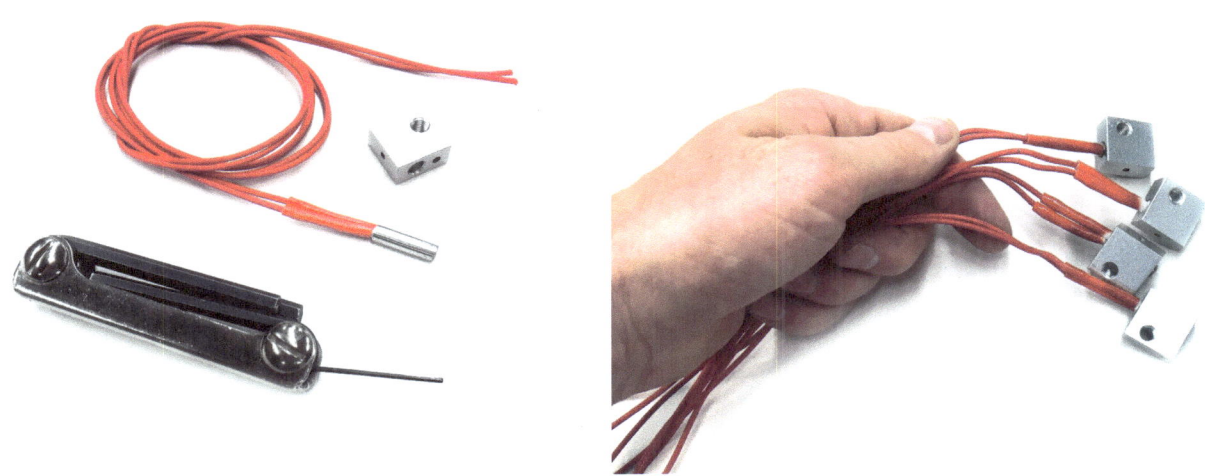

Put a set of four heater elements into aluminum heater blocks. Use a 2mm metric Allen wrench to loosen the set screws of the aluminum heater blocks. Now, fully insert the cartridge heaters as shown, and tighten the set screws.

2.3.2 Bolt heater blocks to pipe flange

An iron pipe flange will act as the oven's radiator. Bolt the heater blocks onto iron pipe flange as shown. Use an M6 stainless machine screw to bolt each block to the underside of the flange. Consider polishing the flange underside to improve surface contact with the heater block.

High temperature thermal grease may be applied but is not necessary. It may help to lengthen the life of the elements, but they are quite inexpensive.

Chapter 2 : Section 3 - Base tray

1.3.3 Mark edges and poke leg-holes through pizza screens

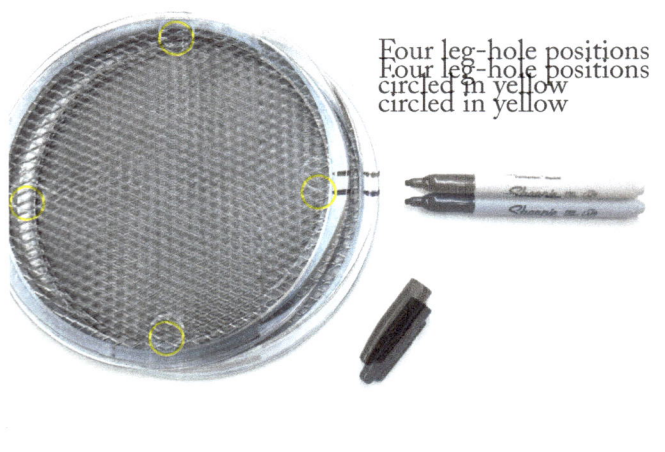

Four leg-hole positions circled in yellow

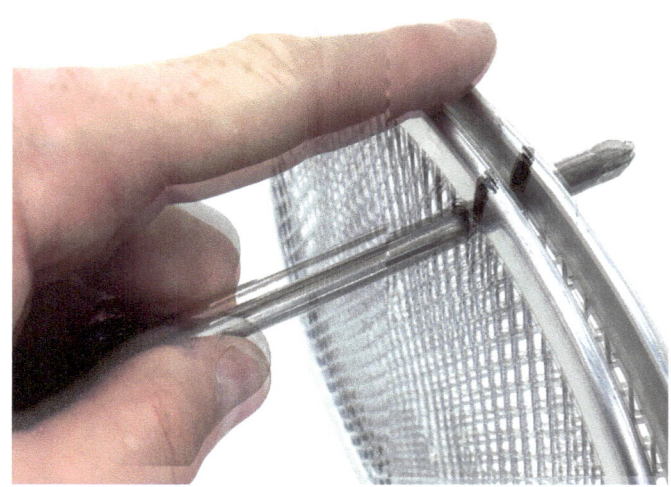

Use a marker to mark the edges of two pizza screens. This will let you line up the holes that you will make. Use a blunt screwdriver to poke four holes through pizza screens in a square pattern. Each hole must be large enough for a 1/4-20 bolt to pass through and will become the position of a leg.

2.3.4 Bolt the pipe flange onto the bottom pizza screen

Choose one screen to be the bottom. Push the bolts from the heater through the middle of the screen. Apply washers and M6 stainless nuts to secure the finished heater block. Tidy up the wires by passing them through the mesh of the pizza screen as shown.

Chapter 2 : Section 3 - Base tray

2:3:5 Install the bolt-legs

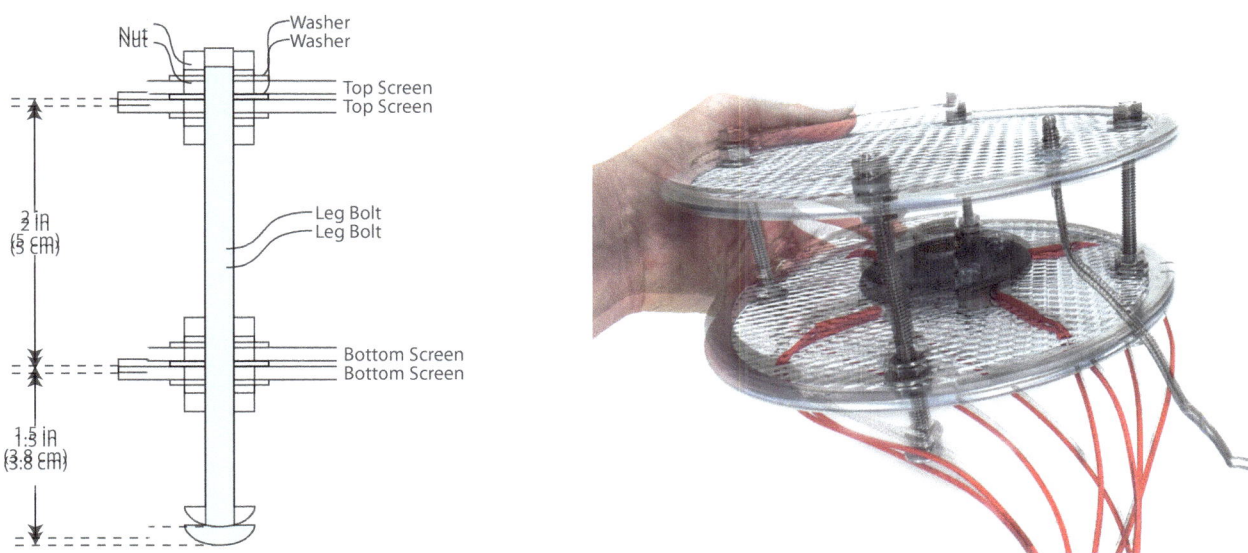

Install the bolt-legs: Use the 1/4-20 x 4 inch (M6 x 10 cm) stainless carriage bolts along with nuts and washers. Space them as shown to leave around 2 inches (5cm) of space between the heater and the top screen. The rounded bolt caps form the feet.

2:3:6 Bolt the thermocouple probe into the top screen

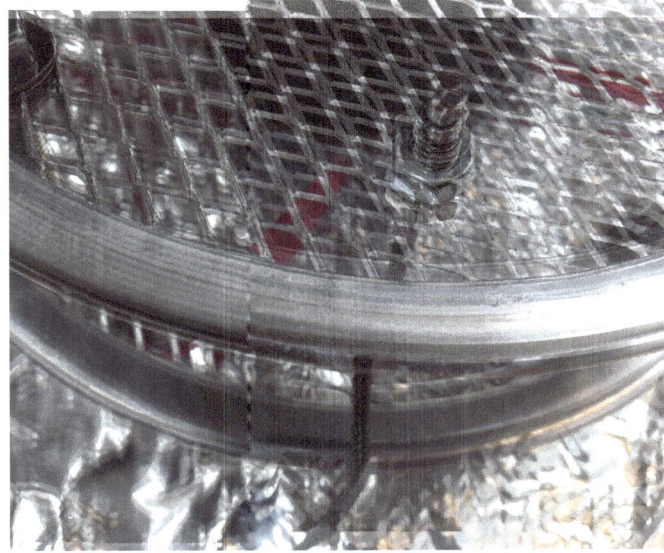

Finally, install the threaded thermocouple probe into the edge of the top rack, and secure with a nut as shown above.

Chapter 2 : Section 3 - Base tray

2.3.7 Place heater rack on the tray, put on the pot, trim the wires

Place the finished heater rack in the center of the pillow; do not apply tape within the autoclave chamber as the tape's adhesive will burn, contaminating medical instruments.

This completes the heater rack section of your autoclave.

Chapter 3 - Build the Electronic Controller

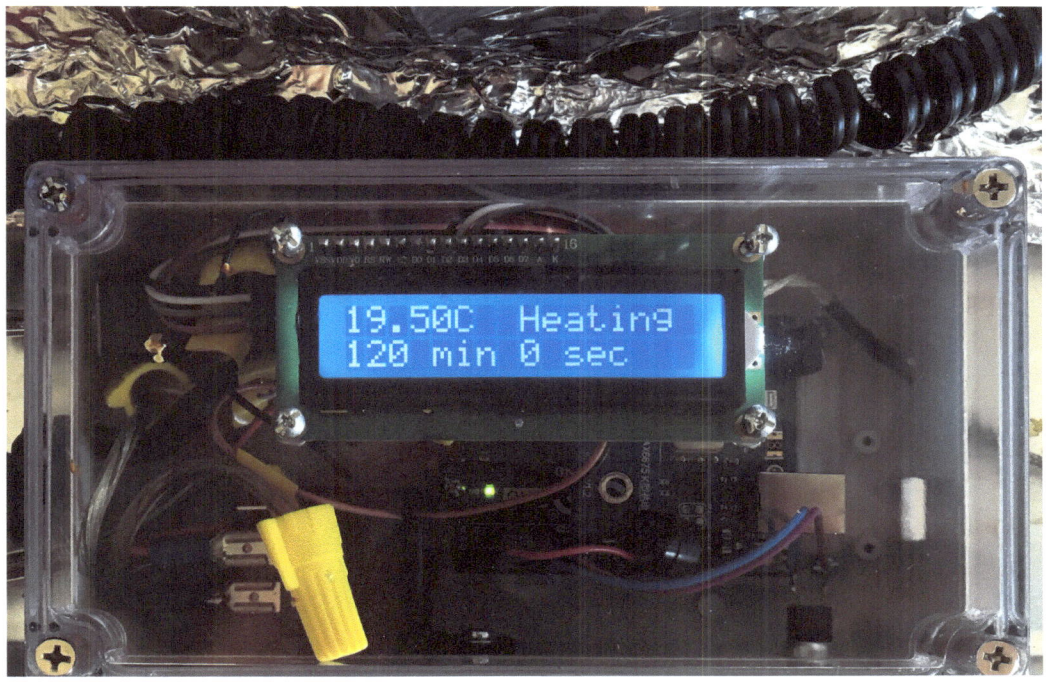

Fig. 3.0 - Close up of the electronic control unit in the process of warming up

3.1 Drill holes in the electronics box
3.2 Install the components
3.3 Do the wiring

Chapter 3 : Section 1 - Drill holes in the electronics box

3.1.1 Print out the drill templates: figures 3.1a and 3.1b

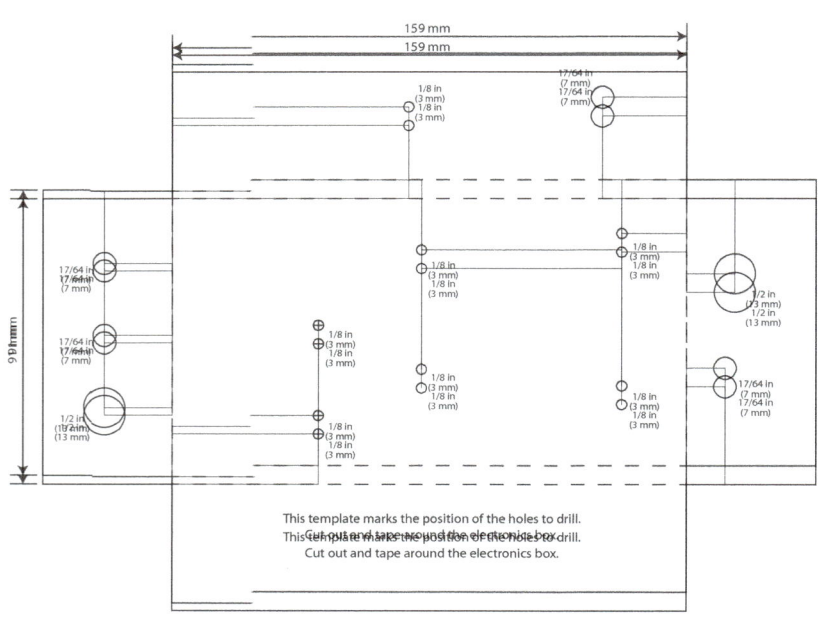

Print out hole-pattern templates, figures 3.1a and 3.1b. The templates are located on the next two pages of this book. They are also in the set of downloaded files from GitHub.

Once you print the pages, you should check the dimensions with a ruler to see if the printout is actually the correct size. You may need to reprint with 'scaling' turned off.

3.1.2 Use template to mark the box for drilling

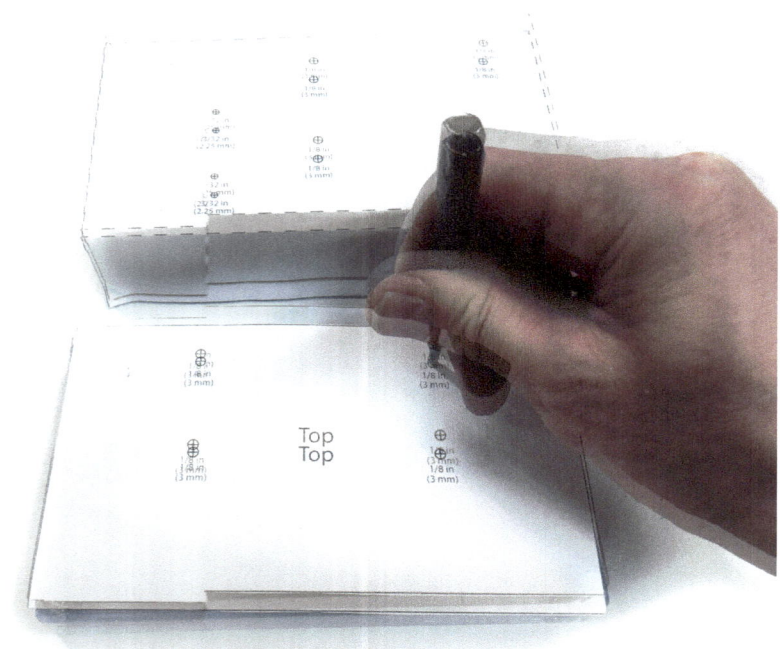

Cut, fold, and tape template to fit over the box as shown. Use a center punch to mark the centers of each hole with a light tap of a hammer.

If you are using a box of a different size, you may cut the template apart into the different faces and apply separately with tape.

Figure 3.1.a - Drill template for the electronics box lid

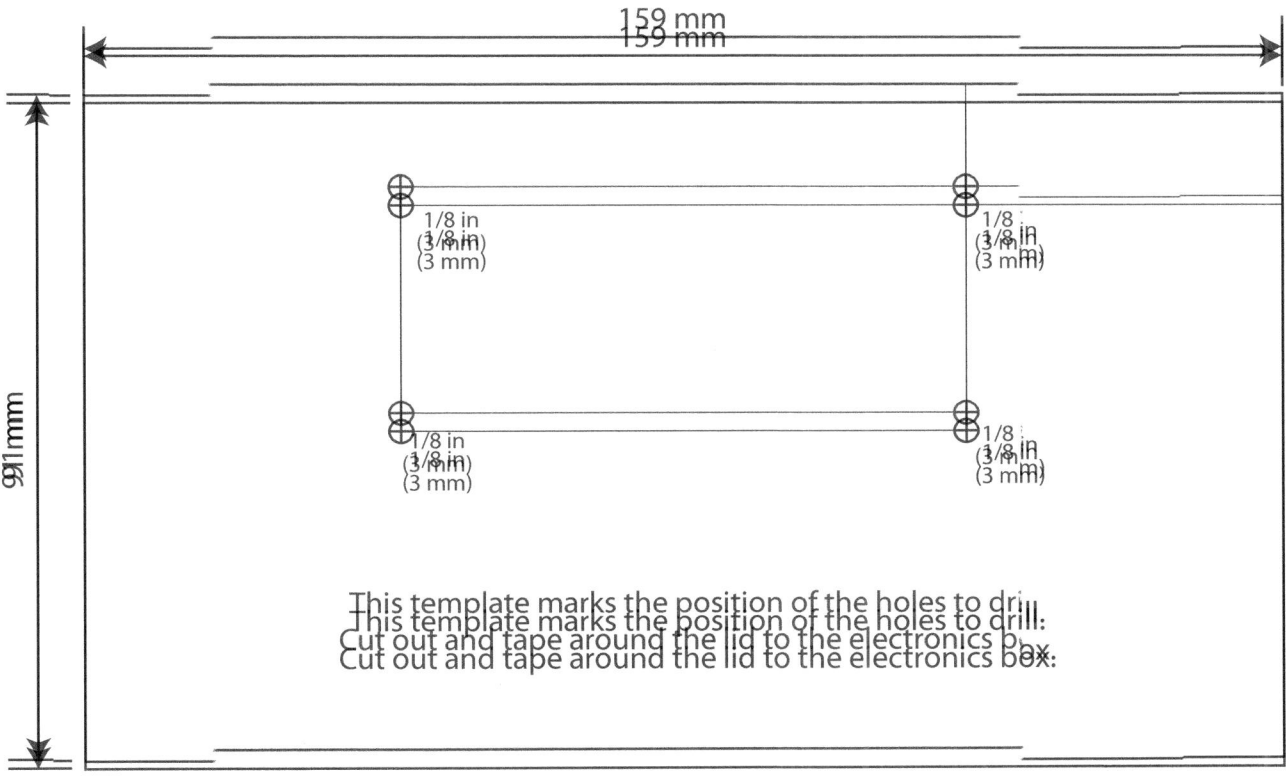

This template marks the position of the holes to drill.
Cut out and tape around the lid to the electronics box.

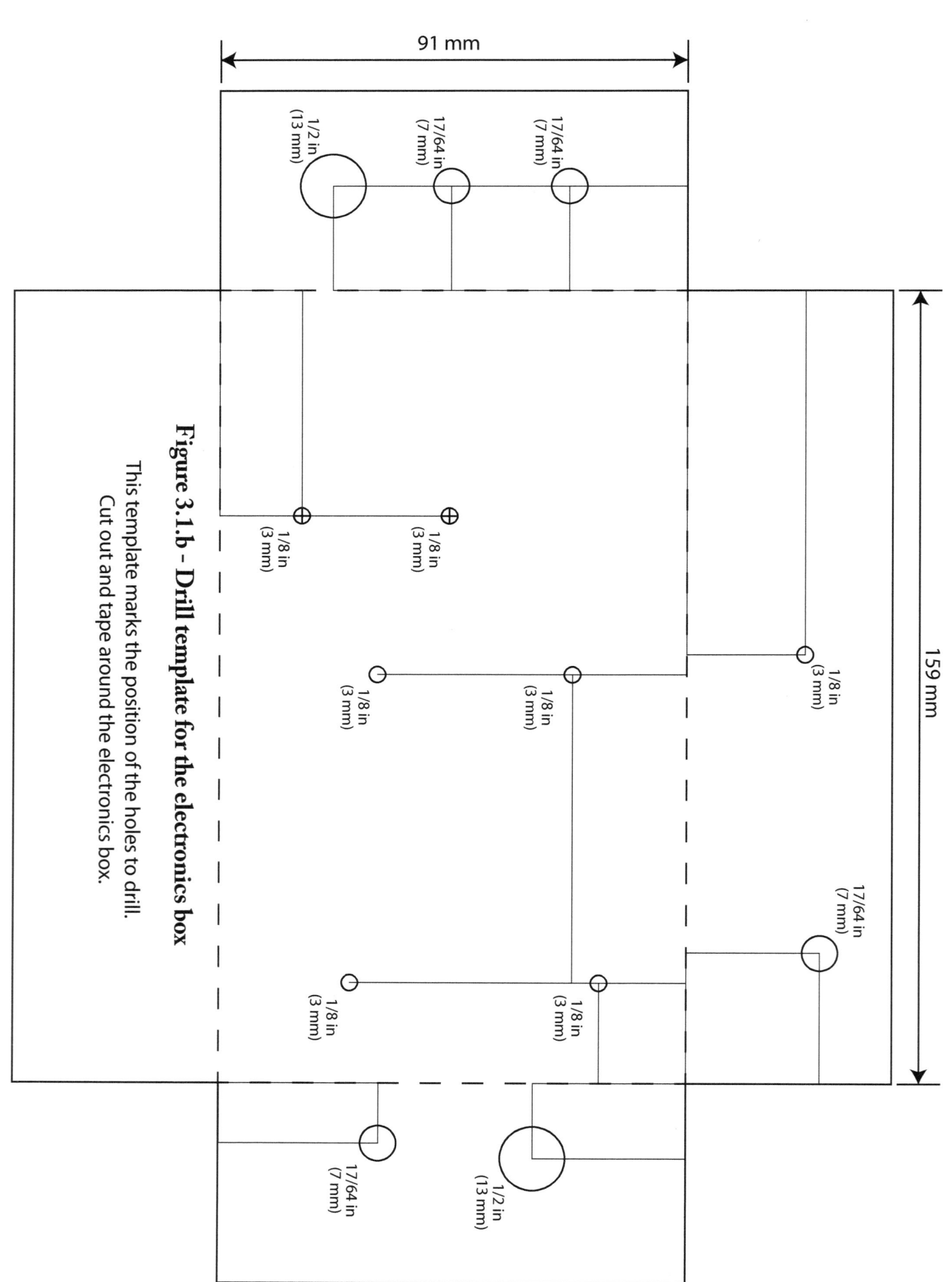

Figure 3.1.b – Drill template for the electronics box

This template marks the position of the holes to drill.
Cut out and tape around the electronics box.

Chapter 3 : Section 1 - Drill holes in the electronics box

3.1.3 Drill the holes using the sizes shown in templates

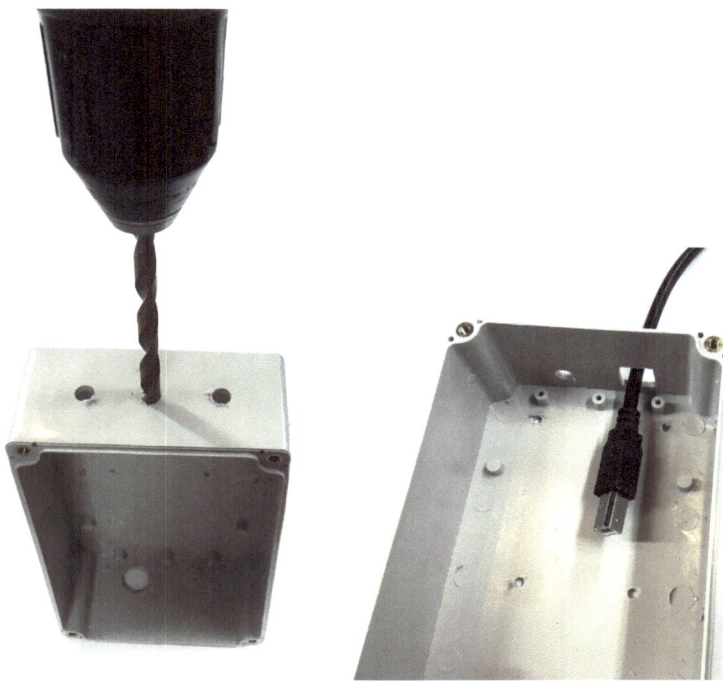

Use the drill with each of the drill bit sizes shown in the template (figures 3.1 a & b) to put holes in the box.

Use a utility knife to square-off the hole marked "USB" so that a printer cable can pass through into the box. (see inset)

3.1.4 Use the box to drill holes through the tray

Select a 1/8-inch drill bit, position the electronics box as shown on the tray.

Now, drill through the four holes labeled "Arduino Bracket Mount Holes" and through the aluminum tray. You may wish to place the tray onto a wooden board or a work bench that can be drilled into.

Open Autoclave: Build an open-source off-grid medical instrument sterilizer by David Hartkop   CC-BY license

Chapter 3 : Section 2 - Install the electronic components

Fig. 3a – Picture of the main components physically installed inside of the electronics box

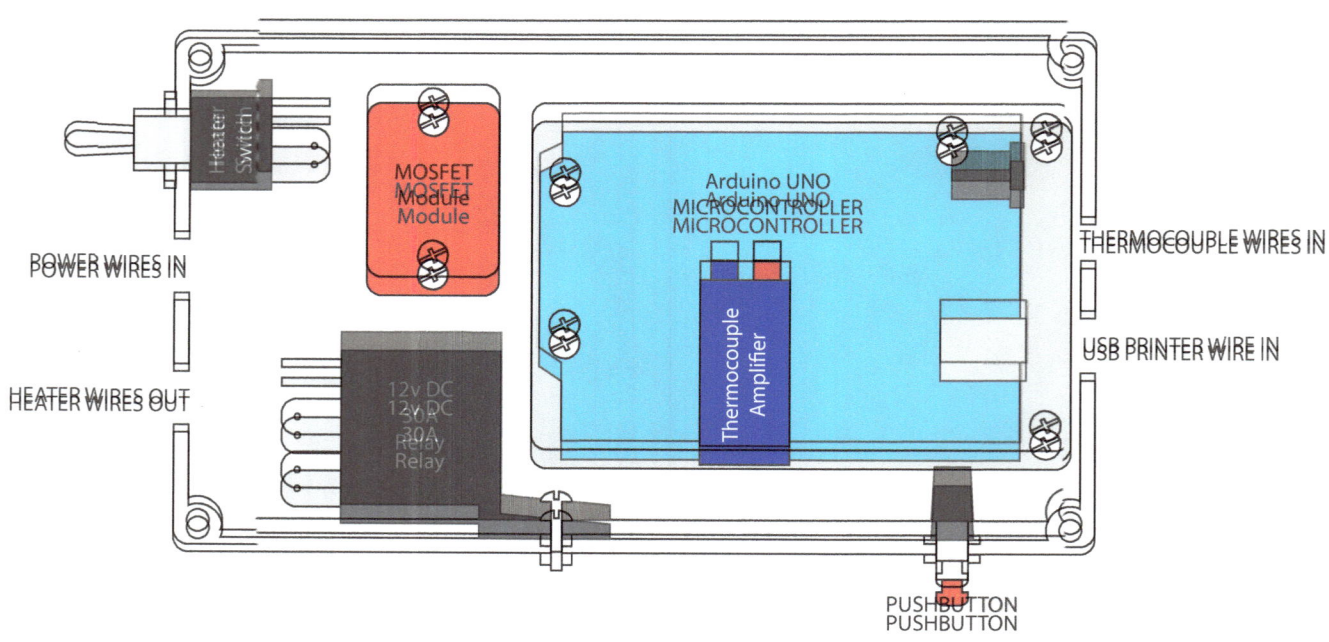

Fig. 3b – Illustration showing the placement of the main components inside the electrical box

Chapter 3 : Section 2 - Install the electronic components

3.2.1 MOSFET mounting screws

The small MOSFET board will be mounted on pair of screws installed up through the bottom of the box. We will use two 4-40 x 3/8 inch (M3 x 10mm) tapered-head screws.

The screws must not make electrical contact with the tray or it causes problems later!

Use a 1/4-inch drill bit to taper the two holes by hand. Insert the tapered screws and secure with nuts from the inside of the box. Finally, insulate the heads of the screws with a strip of electrical tape as shown.

Small self-adhesive rubber feet can also be used to stand the elecrical enclosure a few mm away from the surface of the aluminum sheet.

3.2.2 Install the MOSFET board

Place the small board over the two screws inside the box, and secure with a pair of nuts. Tighten with needle nose pliers. Be careful to avoid touching the pins as much as possible, as they are static sensitive and can be damaged!

Chapter 3 : Section 2 - Install the electronic components

3.2.3 Mount the box onto the tray

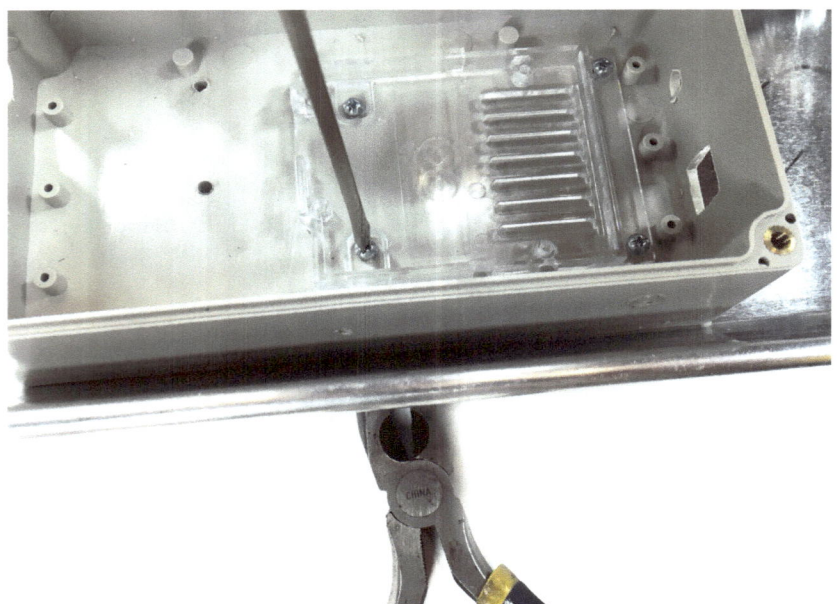

Place the Arduino mounting bracket into the bottom of the box and place the box onto the aluminum tray so that the four mounting holes line up.

Put four of the 4-40 x 3/8-inch (M3 x 10mm) screws down through the four holes as shown. Secure each with a 4-40 (M3) nut on the bottom of the tray. Tighten using a screwdriver and needle nose pliers.

3.2.4 Bolt the Arduino Uno R3 to the mounting bracket

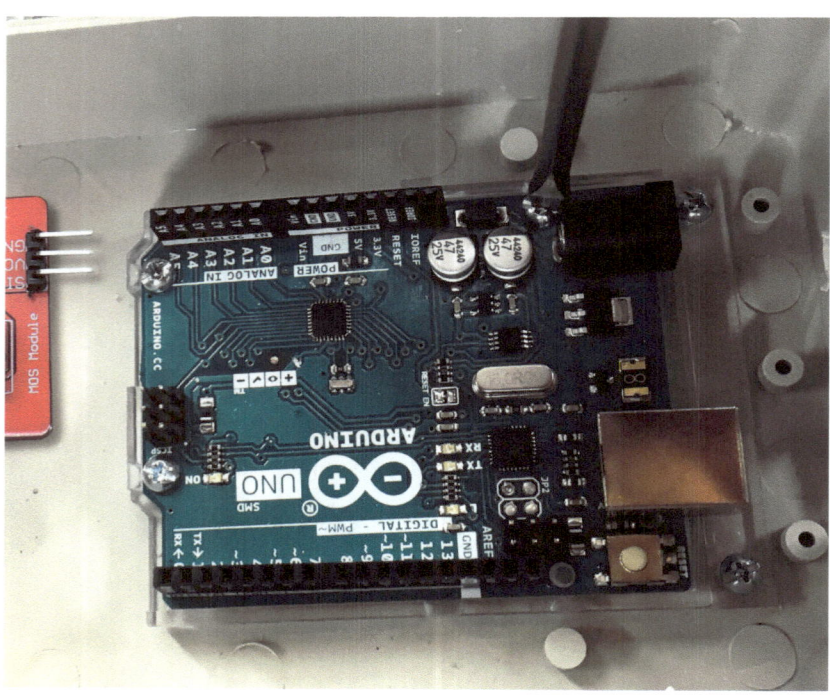

Bolt the Arduino Uno R3 microcontroller into the mounting bracket using three 4-40x1/4-inch (M3 x 6mm) screws. Note that one hole in the Arduino is left unused because it is too tight a fit for the screw head.

Chapter 3 : Section 2 - Install the electronic components

3.2.5 Put wire on the automotive relay

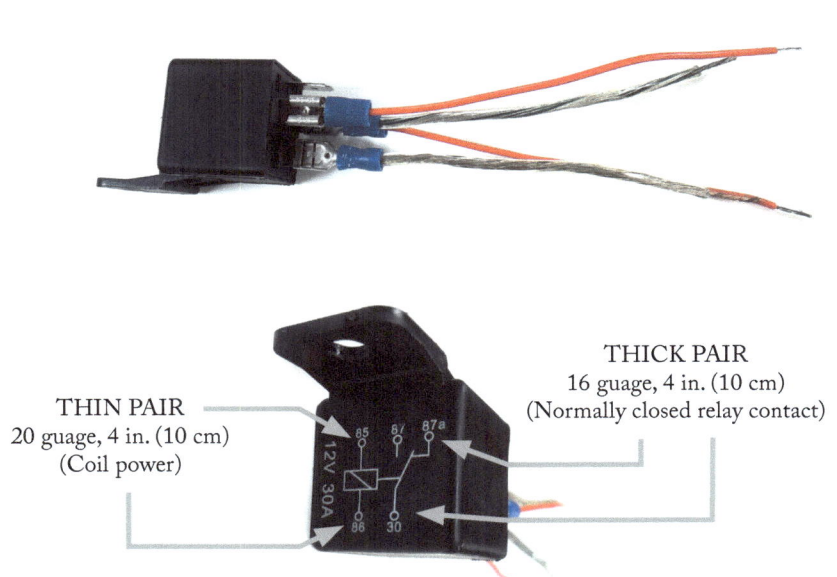

Locate four female blade type crimp connectors. Now, use the wire strippers and crimpers to make two pairs of wires with connectors attached:

Thin pair

(26 gauge, 4 inches long)

Thick pair

(16 gauge, 4 inches long)

Connect the thin pair of wires to the two leads for the coil.

Connect the thick pair of wires to the C and N.C. pins of the relay.

See image and insets for details.

3.2.6 Mount the automotive relay into the box

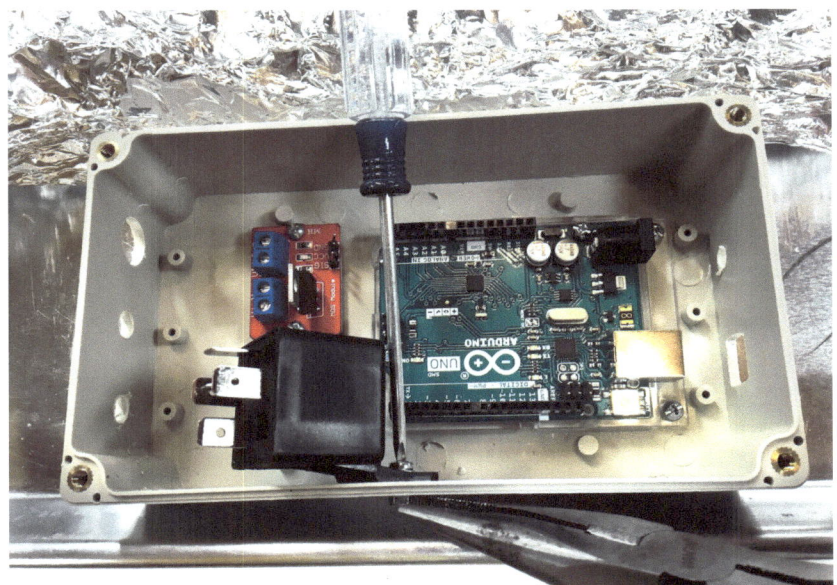

Fig. 3.2.6 – Mounting automotive relay. Wires removed for clarity of picture.

Mount the Automotive Relay using a 4-40 x 3/8-inch (M3 x 10mm) screw, a nut, and a washer as shown in the picture.

You can tilt the relay, so its pins face up toward you for now. It will be tilted down with the pins to the left to close the box.

Chapter 3 : Section 2 - Install the electronic components

3.2.7 Put wires onto the heater power switch

Prepare the heater power switch. Attach a pair of 4 inch (10 cm) long 16-gauge wires to the two leads of the switch. You may insulate the leads with shrink wrap tube if you wish.

3.2.8 Mount the power switch into box

Install the heater power switch through the hole as shown in the figure. Tighten its ring nut with needle nose pliers.

Chapter 3 : Section 2 - Install the electronic components

3.2.9 Put wires onto the push button

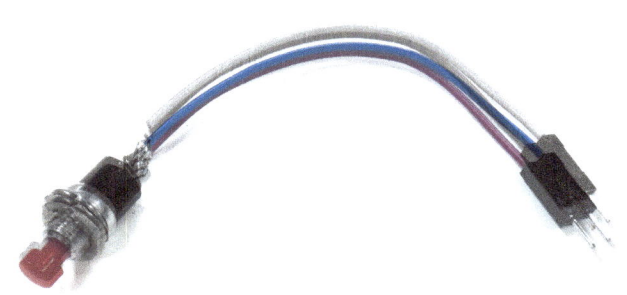

Attach a pair of 4-inch (10 cm) pin connector hook-up wires to the two leads of the small pushbutton as shown.

You may need to use a soldering iron or crimp connectors. You may also cover the leads of the pushbutton with shrink wrap tubing if you wish.

3.2.10 Install the push button

Install the push button into the box and tighten its ring with needle nose pliers.

Chapter 3 : Section 2 - Install the electronic components

3.2.11 Put wires onto a small PC speaker

Prepare the PC speaker. Attach a pair of 2-inch (5 cm) pin connector hook up wires to the two leads of the small PC speaker. The wires on these tiny speakers tend to be fragile. You might wish to reinforce with hot-melt glue or electrical tape before installing.

3.2.12 Connect the speaker to the Arduino Uno

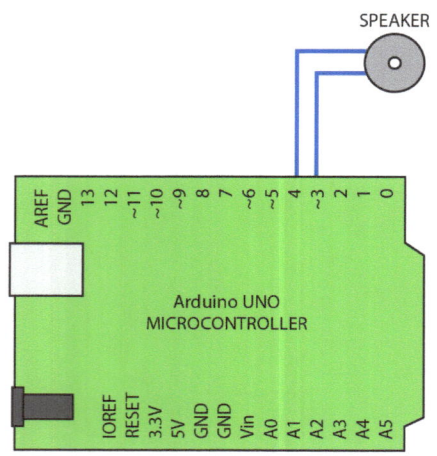

Install the mini PC speaker by plugging its leads into pins on the Arduino Uno board, as shown in figure.

3.2.13 Connect the control button to the Arduino Uno

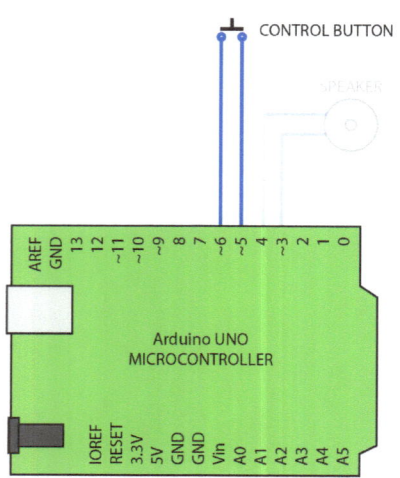

Install the mini control button by plugging its leads into pins on the Arduino Uno board, as shown in figure.

Chapter 3 : Section 2 - Install the electronic components

3.2.14 Thermocouple probe wires into the box

Locate the thermocouple probe and thermocouple amplifier module.

Run the probe's leads into the box through the hole as shown. Use a screwdriver to connect the probe's leads to the amplifier Red to + and Blue to - as shown.

3.2.15 Install the thermocouple amplifier

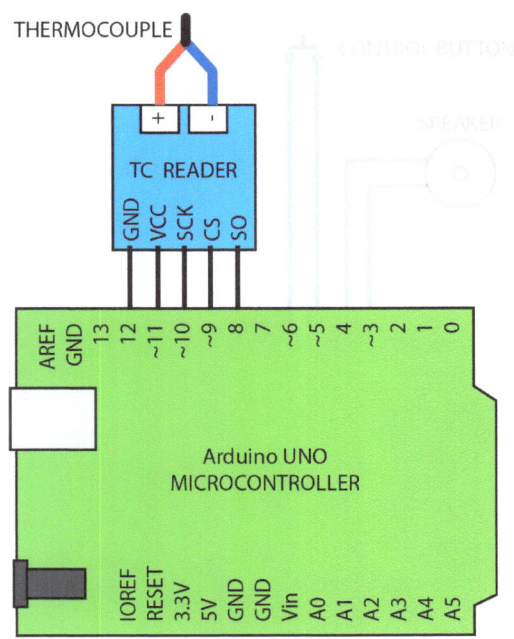

Bend the five leads of the small amplifier board so that they stand directly out from the board.

Now, the amplifier's 5 pins can be pushed into the corresponding holes in the Arduino board.

Double check that you have inserted the Thermocouple Amplifier pins into the Arduino Uno board as shown in the schematic.

Chapter 3 : Section 2 - Install the electronic components

3.2.16 Put wires onto the LCD screen module

Find four jumper wires of the 4-inch length. Attach them to the four leads of the LCD. The connections to the LCD tend to be very sensitive, so you may wish to secure the small push-on pin connectors to the LCD module with hot melt glue or even superglue.

3.2.17 LCD screen into the lid of the box

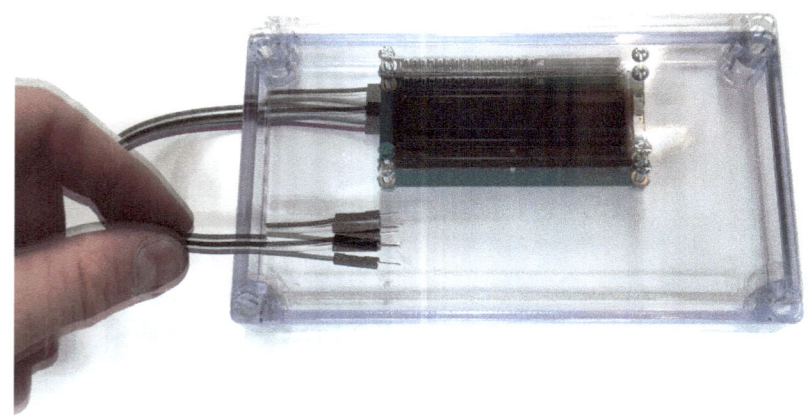

Install the LCD module onto the transparent lid of the box as shown using 4-40 x 3/4-inch (M3 x 20 mm) screws. Secure the LCD module by tightening a pair of 4-40 (M3) nuts on each screw. Use a screwdriver and needle nose pliers.

Chapter 3 : Section 3 - Finish the wiring

3.3.1 Connect the LCD screen to the Arduino

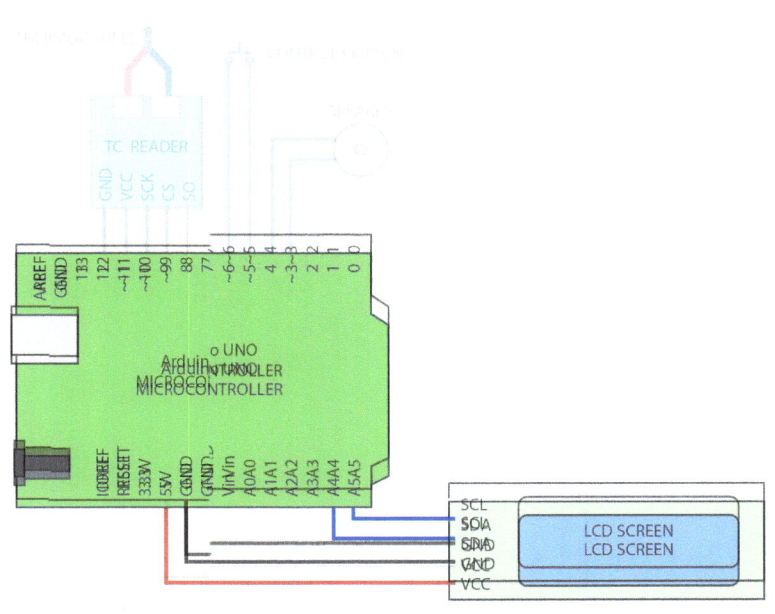

Connect the LCD to the Arduino Board. Find the four wires from the LCD module, and connect them to the Arduino board as shown in the figure. Double check that the pins are connected correctly.

3.3.2 Connect the MOSFET to the Arduino

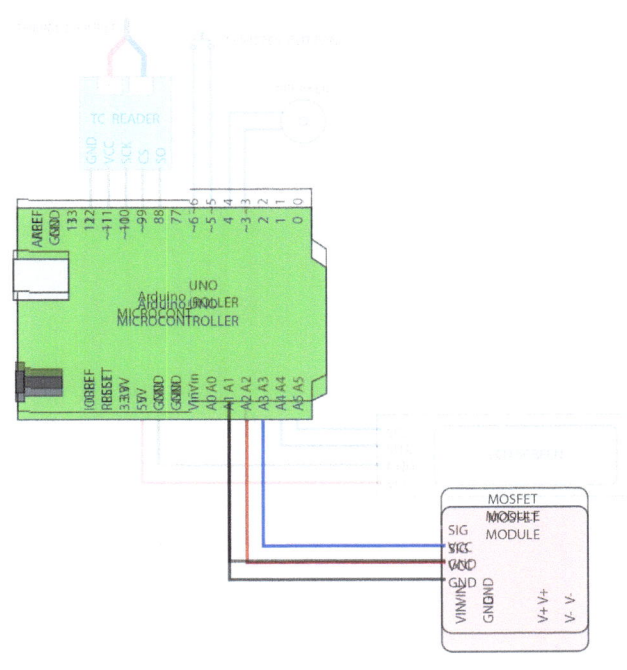

Connect the MOSFET module to the Arduino Board. Find three of the pin connector wires and connect them to the Arduino pins as shown in the figure.

This part is static sensitive, so be careful to avoid touching its pins with your fingers.

Chapter 3 : Section 3 - Finish the wiring

3.3.3 Insert power wire cables into box

Insert the two wires from the 12VDC power cable into the hole in the box as shown in the figure. Use white electrical tape and a black marker to label the wires (+) and (-). If you are not sure which wire is which, you can use a volt-ohm meter to check the continuity.

For a cigarette lighter plug, the (+) lead is always the very tip of the plug. See details about using a volt-ohm meter in section 5.2 on page 69.

3.3.4 Run heater-power wires out of box

Run a pair of 8 inch (20 cm) long 16-gauge wires into the hole in the box as shown. These will be the wires that carry power out to the heater coils. Mark the wires (+) and (-) on the ends inside the box.

Chapter 3 : Section 3 - Finish the wiring

3.3.5 Wire the positive (+) power wires

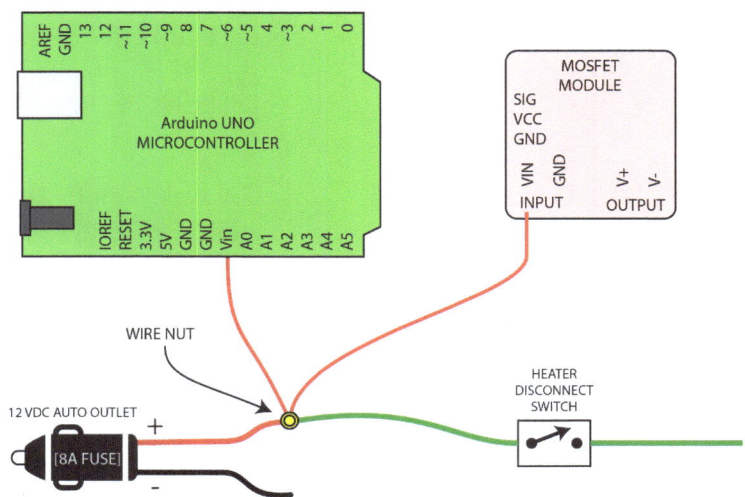

Wire the positive (+) power wire. For this step, we will use a small wire nut to connect together the four wires that are meant to be positive, as shown in the figure.

Use one wire nut to connect the following:

(+) 12 Volt Power Wire

(+) power input for Arduino Uno Microcontroller

(+) power input for the MOSFET module

One lead of the heater disconnect switch

3.3.6 Wire the negative (-) power wires

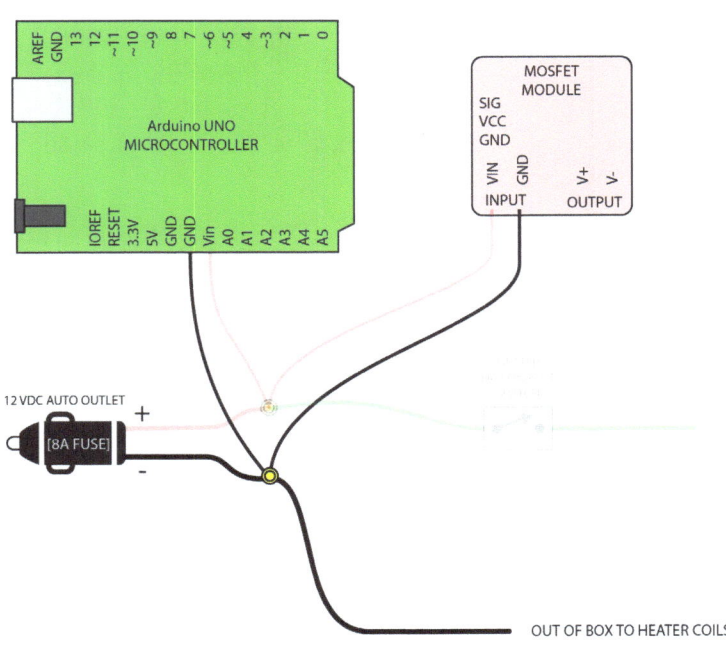

Wire the negative (-) power wire. For this step, we will use a small wire nut to connect together the four wires that are meant to be negative, as shown in the figure.

Use one wire nut to connect the following:

(-) 12 Volt Power Wire

(-) power input for Arduino Uno Microcontroller

(-) power input for the MOSFET module

(-) power output wire to the heater coils

Chapter 3 : Section 3 - Finish the wiring

3.3.7 Connect MOSFET output to Relay's coil wires

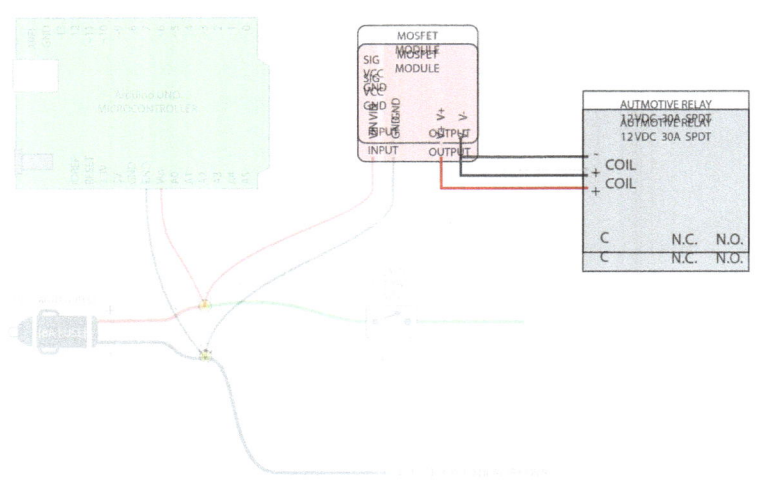

Use a pair of thinner (26 gauge is ok) hookup wires to connect the two output wires from the MOSFET module to the two COIL input wires of the automotive relay.

(The +/- polarity does not actually matter because the relay's coil will work fine either way.)

3.3.8 Connect the relay contact wires to the switch and to the heater coils

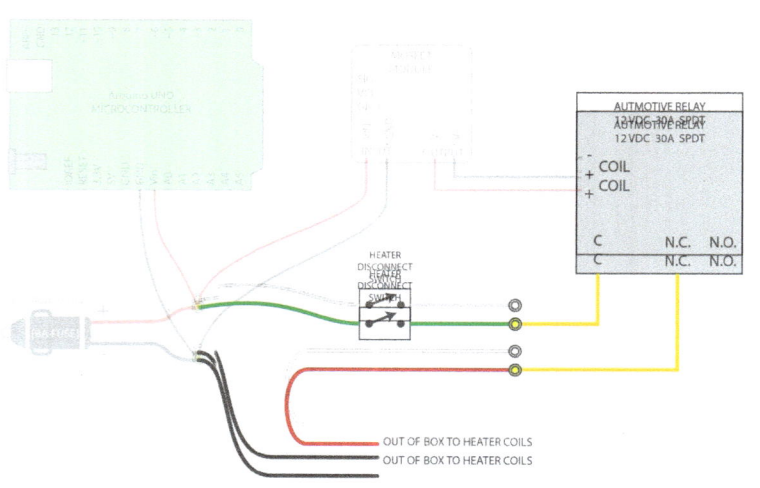

For this step, we will be concerned with the two thicker wires coming from the relay. The two thicker wires should be connected to the C and the N.C. contact pins of the relay, as shown in step 2.2.5.

Use a small wire nut to connect one of the relay contact wires to the unused wire from the heater disconnect switch.

Use a small wire nut to connect the other thick wire from the automotive relay to the positive power output wire that goes out of the box to the heater coils.

Chapter 3 : Section 3 - Finish the wiring

3.3.9 Tidy the wires and close the box

Rotate the automotive relay down. Carefully tuck the connected wires down into the box around the edges.

Use some zip ties to tidy up the wire and secure them within the electronics enclosure.

Finally, close the box, taking care that the back of the LCD screen has enough room. Bend the thick wires down or back as needed inside the box.

3.3.10 Wire the heater elements to the power output wires

Wire the heater elements to the Power Output Wires as shown in the figure. One pair of elements is wired in series, one pair is wired in parallel, and then both pairs are wired in parallel.

Note: Each 12v 40W heater cartridge has a resistance of 3.875 ohms. The wiring pattern shown in below has a total resistance of 1.55 ohms. At 12 volts this draws 7.75 amps for a total heating power of 93 Watts.

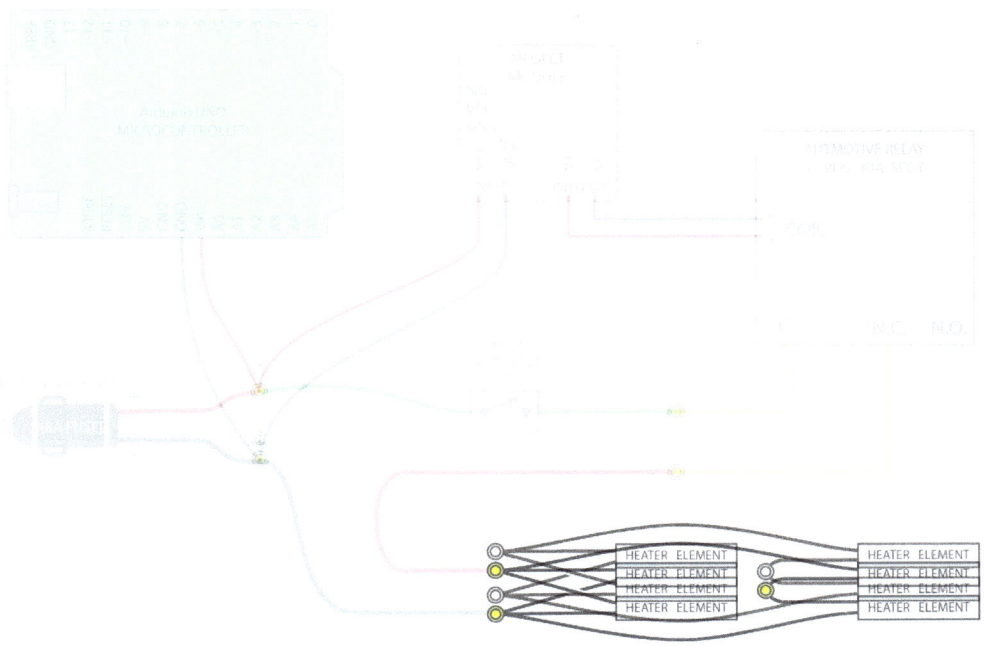

Chapter 3 : Section 3 - Finish the wiring

3.3.11 Double check power polarity with a volt-ohm meter

If the power wires are connected backwards, the Arduino Uno microcontroller or to the MOSFET, it will most likely burn these parts out.

Double check that you have connected the power polarity (+ vs -) to the proper pins of the arduino.

Double check that you have connected the power polarity (+ vs -) to the proper pins of the MOSFET module.

See the diagram for how a Volt-Ohm meter can be used to check this.

The schematic on the following page shows the Arduino control and power system in its entirety, along with the list of parts and descriptions.

Chapter 3 : Section 3 - Finish the wiring

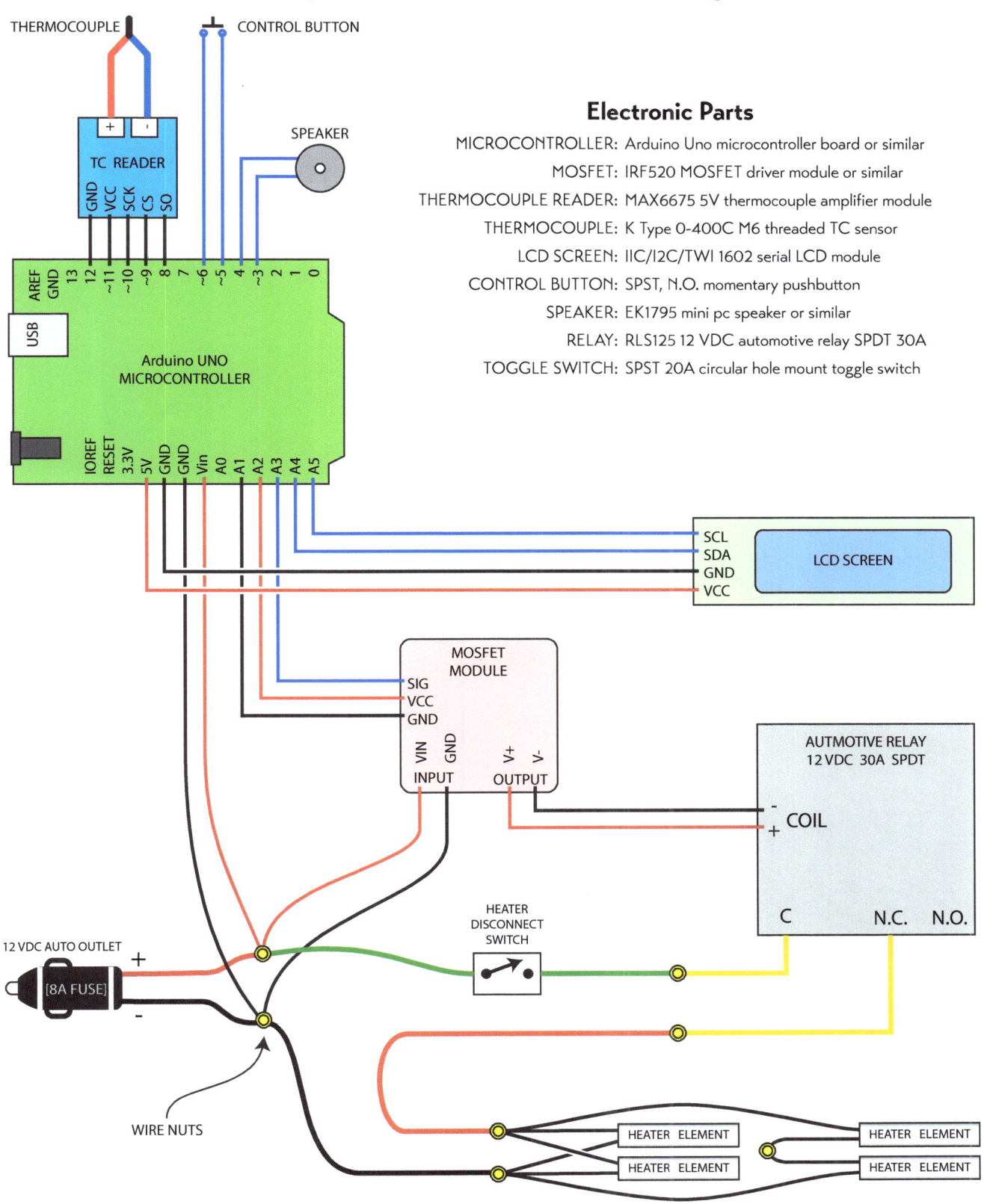

Open Autoclave: Build an open-source off-grid medical instrument sterilizer by David Hartkop CC-BY license

Chapter 4 - Set Up The Control Computer Software

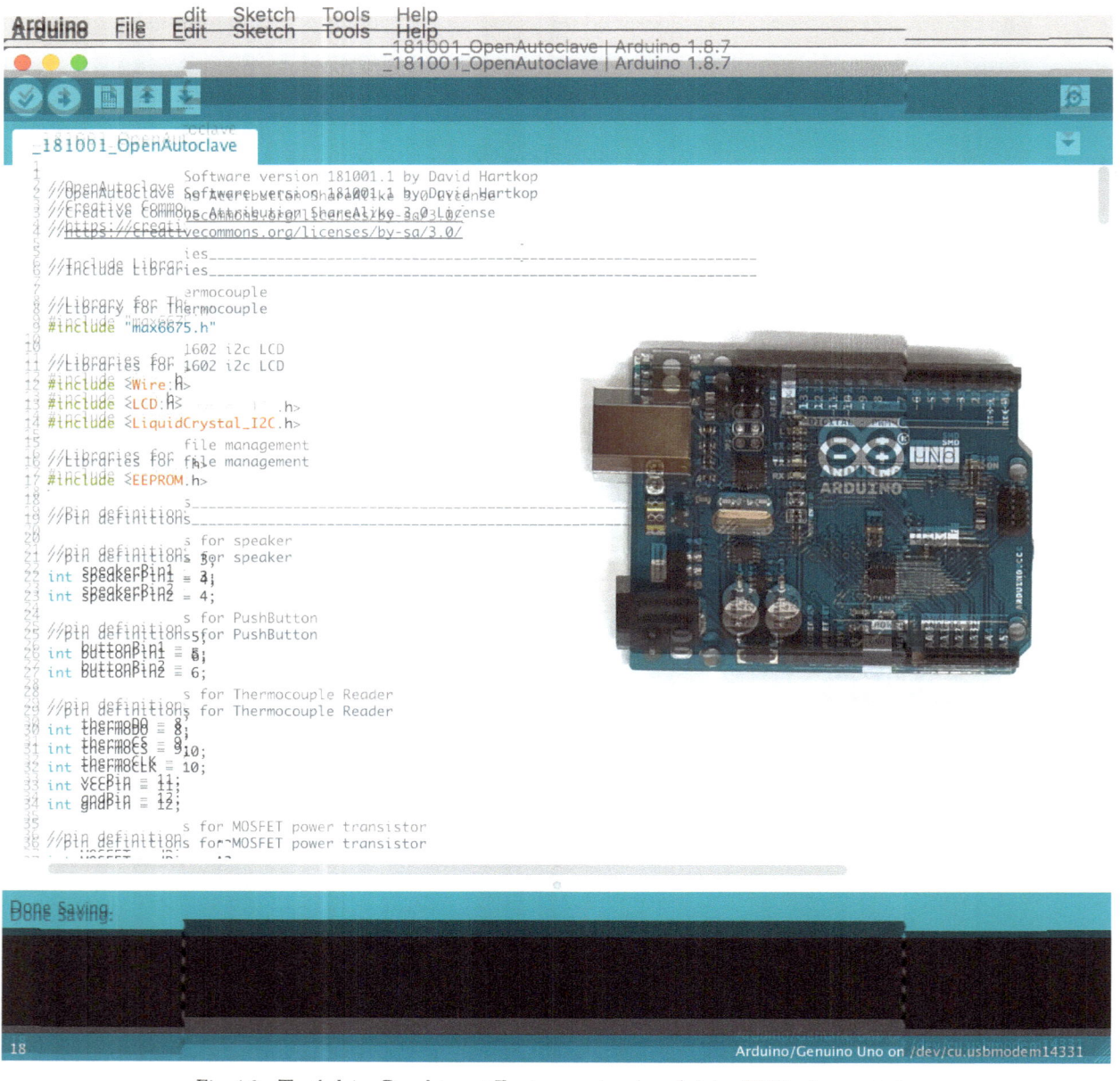

Fig. 4.0 - The Arduino Development Environment and an Arduino UNO micro-controller

4.1 Install the Arduino software

4.2 Download the OpenAutoclave software

4.3 Upload the software to the micro-controller

Chapter 4 : Section 1 - Install the Arduino Software

4.1.1 Download the free Arduino software to your computer

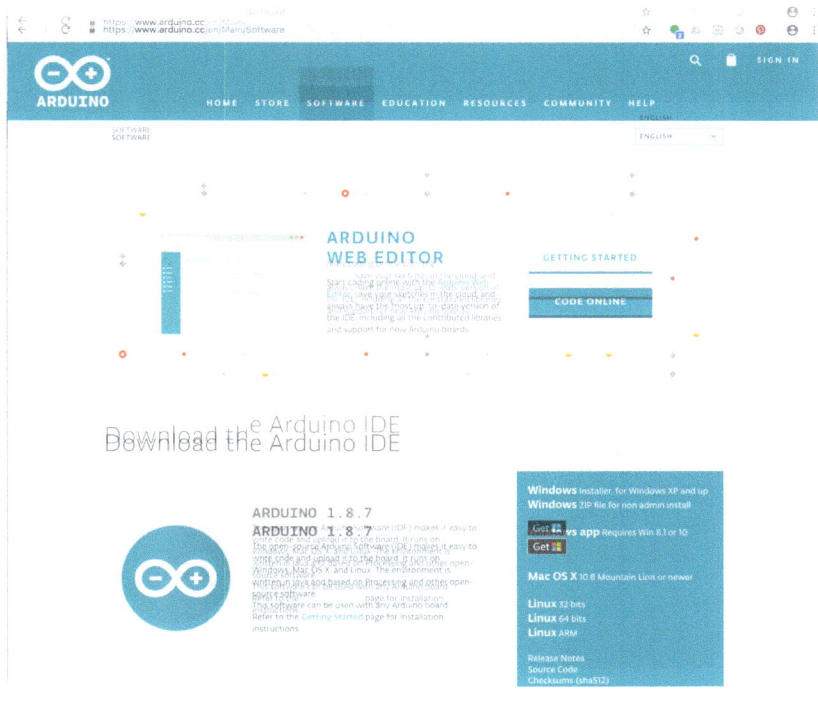

You will need a laptop or desktop computer connected to the internet. Go to this web address:

www.arduino.cc/en/Main/Software

Choose the download that matches your computer operating system, for instance, "Windows Installer" for windows. This software will let your computer connect to and program the Arduino microcontroller.

4.1.2 Unzip the downloaded Arduino software

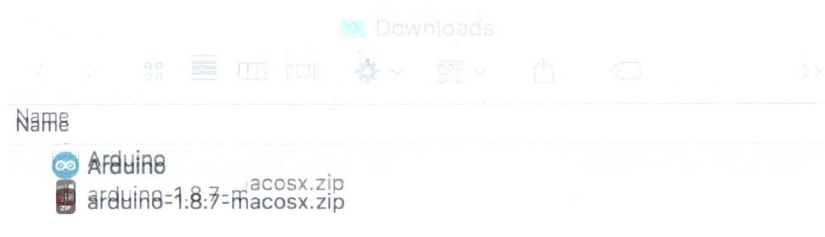

Once the software has downloaded, find it in the downloads directory, or by double clicking on a link that may pop-up in your web browser.

The file will be in .zip format at first. Right click on it and choose to "Extract all" or double click on the .zip on a Mac. This will turn it into an ordinary folder.

Chapter 4 : Section 1 - Install the Arduino Software

4.1.3 Install the Arduino IDE software on your computer

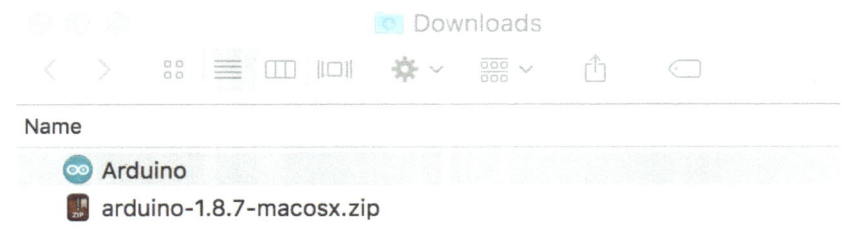

Within the folder, there should be an installer application. Double click on this and follow the steps to install the software on your computer.

On a Mac, this step just requires dragging the Arduino icon from the downloads folder into the application folder.

4.1.4 Open the Arduino IDE software

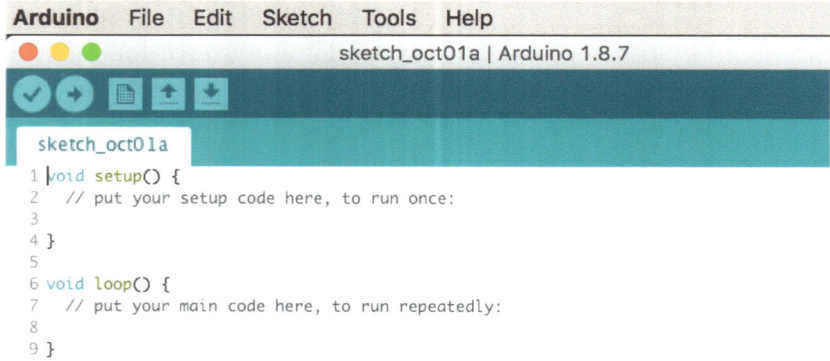

Open the Arduino software by double clicking on the icon in your "Applications" directory. You may need to do a quick search for Arduino on your computer to find it.

If everything is working, the software will start and open with a blank program window or "sketch" as shown in the picture.

If the Arduino software has difficulty starting, you may need to install a new driver on your system. This is a common issue with Windows 10. Perform a quick Google search for "Windows 10 driver for Arduino" and find some current instructions, as fixes for this are subject to change.

Chapter 4 : Section 2 - Download the OpenAutoclave Software

4.2.1 Download the Open Autoclave Software from GitHub

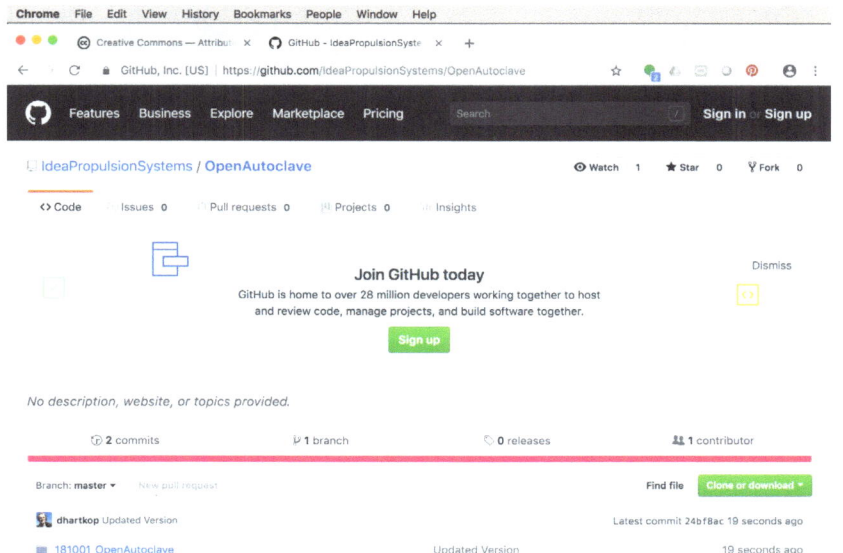

Download the Open Autoclave software from the website GitHub. GitHub is an online repository of shared software. It makes it easy for developers like me to share software projects with others. Go to the following address:

**https://github.com/
IdeaPropulsionSystems/
OpenAutoclave**

See the last page of this book, "Links to Online Resources" for more locations to download the software.

4.2.2 Choose the option to "Download ZIP" file

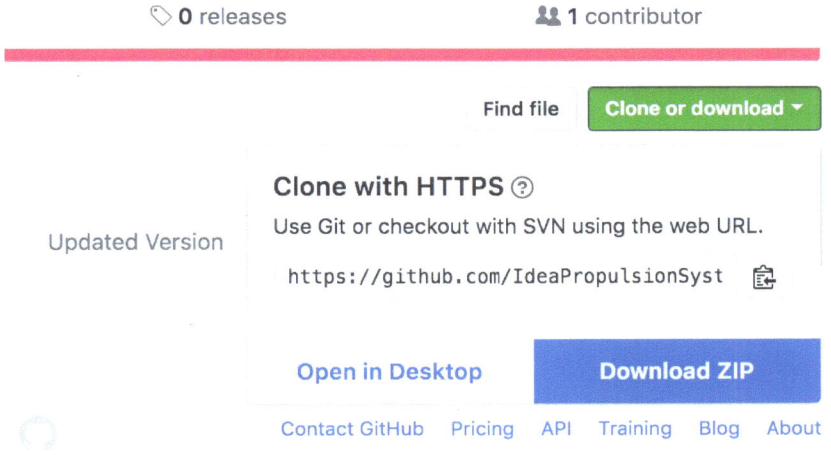

GitHub should load to show the download page for the project. Click on the button marked "Clone or download" and select "Download ZIP." This should download a zip file to the "Downloads" directory on your computer.

Chapter 4 : Section 2 - Download the OpenAutoclave Software

4.2.3 Put the OpenAutoclave software in your Documents folder

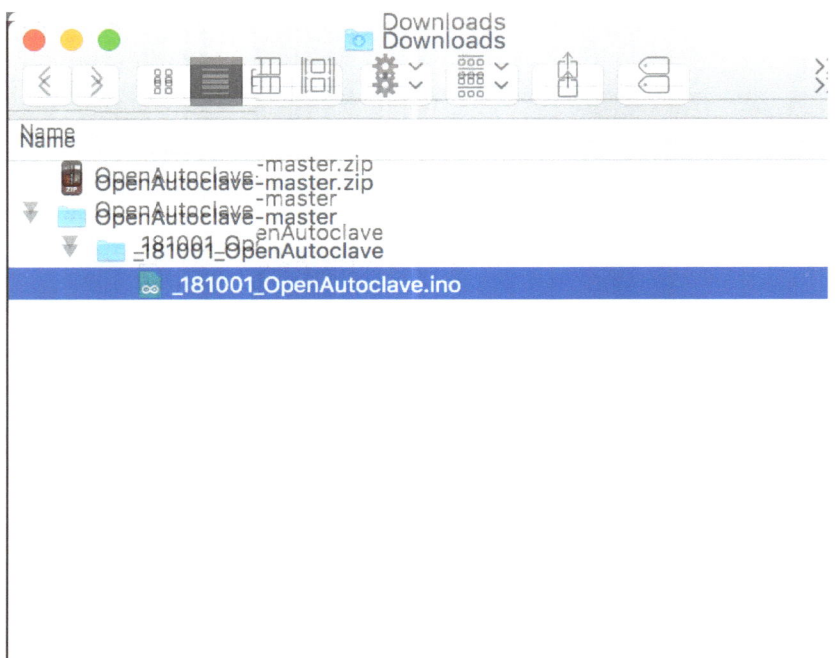

Locate the downloaded ZIP file on your computer. It should be in your "Downloads" directory. Unzip the file by right clicking and choose "extract all" or by double clicking on a Mac.

This should produce a folder named "OpenAutoclave" containing a few files as shown in the picture.

You should move this folder to somewhere easy for you to find. The desktop? The Arduino directory inside of "Documents"? Both good options.

4.2.4 Open the Arduino IDE software

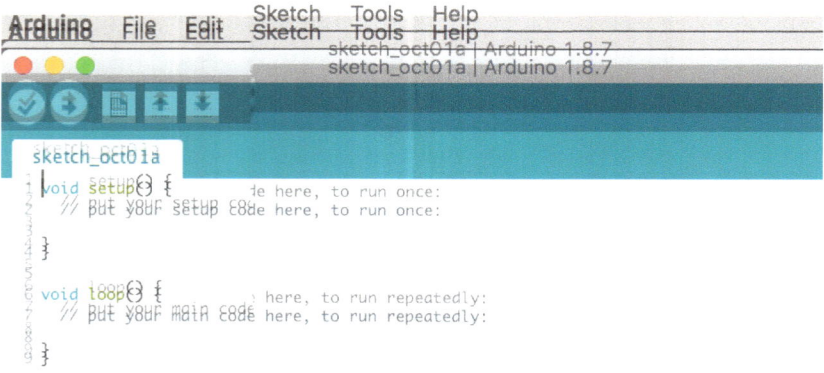

Open the Arduino software by double clicking on the icon in your applications directory. You may need to do a quick search for Arduino on your computer to find it.

If everything is working, the software will start and open with a blank program window or "sketch" as shown in the picture.

If the Arduino software has difficulty starting, you may need to install a new driver on your system. This is a common issue with Windows 10. Perform a quick Google search for "Windows 10 driver for Arduino" and find some current instructions, as fixes for this are subject to change.

Chapter 4 : Section 3 - Upload the software to the Arduino

4.3.1 Prepare to upload software to the Arduino

Switch the Heater Power switch to the "Off" position on the electronics box.

Now, connect your computer to the Arduino using a USB printer cable. The USB will power the computer during the next steps.

4.3.2 Open the Open Autoclave software in the Arduino IDE

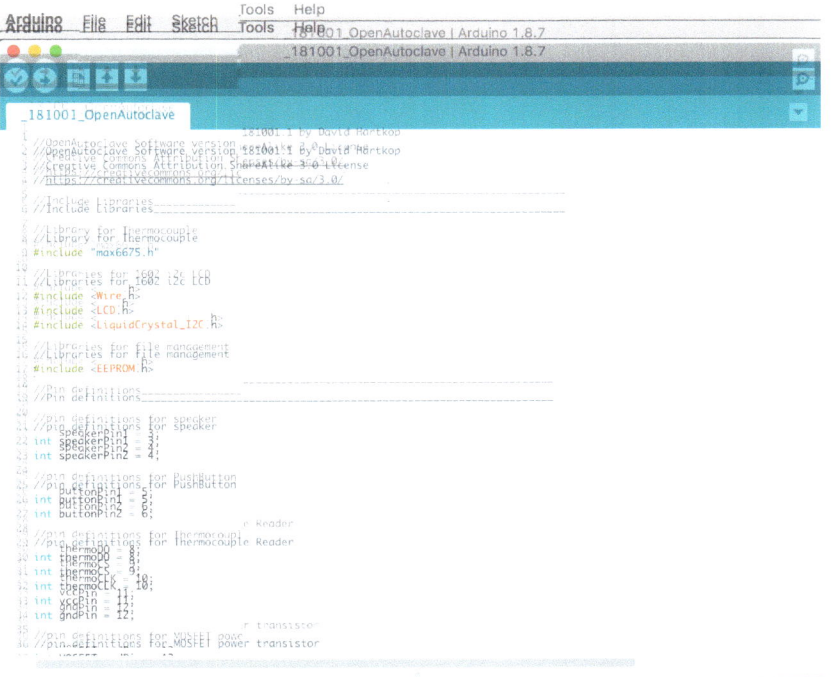

First, launch the Arduino IDE software.

Next, go to File / Open... and find the file named as follows:

_OpenAutoclave_v181001.ino

(It will be in that folder from GitHub.)

Open it. It will open a window of code, as shown at left.

Note: The number "_v181001" may be different because it is a newer version. Just use the most recent version of the file.

Chapter 4 : Section 3 - Upload the software to the Arduino

4.3.3 Select "Arduino/Genuino Uno" from the list

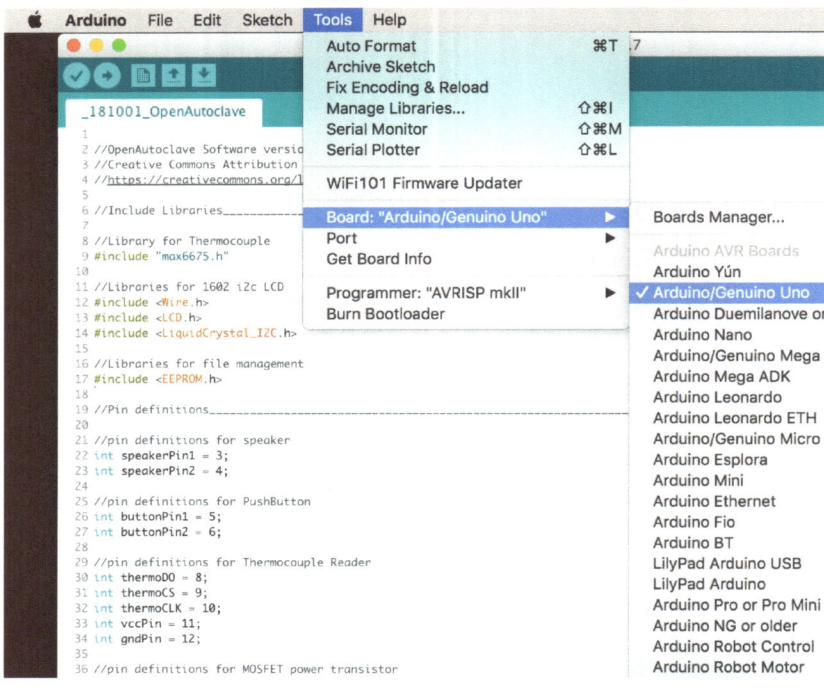

In the Arduino program, go to Tools / Board

From the fly-out list, choose "Arduino/Genuino Uno" from the fly out list.

This tells the Arduino IDE program what board you are using and compiles the code accordingly.

4.3.4 Set the Port location and Upload the software

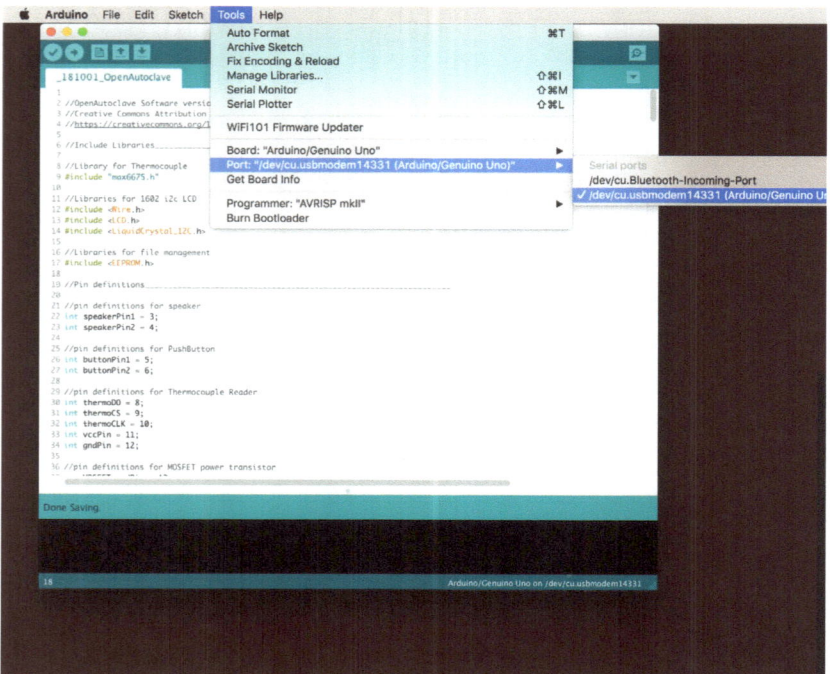

In the Arduino program, from the drop down menus at the top, select Tools / Port.

From the fly-out list, choose one that has ".usbmodem" in the name. One may even say "Arduino UNO" in the name - if so, choose that one.

Now, in the Arduino program, click on the small arrow button at the top of the window. This will begin the software upload to the microcomputer. A message near the bottom will display the message "Uploading…"

Chapter 4 : Section 3 - Upload the software to the Arduino

4.3.5 Prepare to upload software to the Arduino

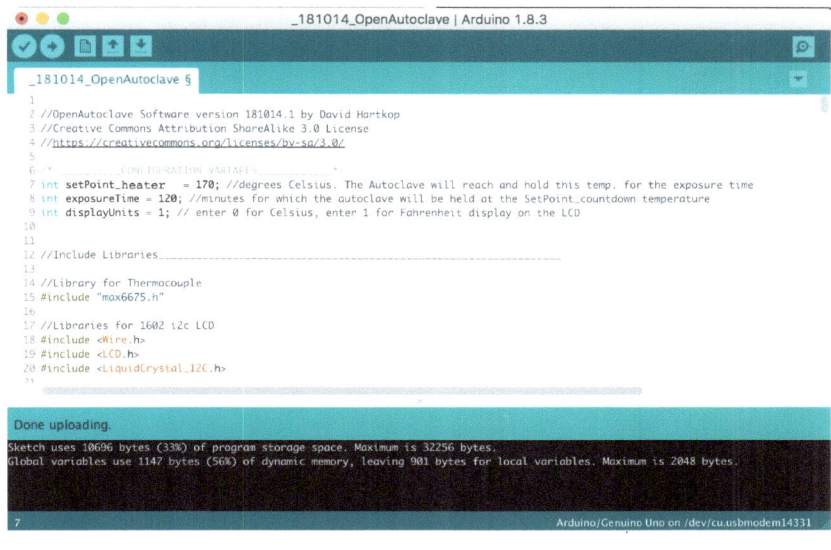

In a perfect world, the message will change from "Uploading…" to "Done Uploading."

If that doesn't happen, don't worry. It probably means you need to go back to step 3.3.4 and try selecting a different port that has ".usbmodem" in the name.

If none appears, you may need to try unplugging the USB connection to the Arduino, then plugging it back in.

If a ".usbmodem" port still does not appear, you may need to install some Arduino drivers for your computer. Do a Google search for "Arduino drivers for Windows 10" etc. (The fixes for issues like this change frequently over time, so you're best off researching yourself as needed.)

4.3.6 Software successfully installed

Once the software is successfully installed into the Arduino, you should hear a cute little "start sound" and the display screen should light up to display the current temperature.

If you are still having difficulty getting to this point, don't worry.

Go to section 4.4 Troubleshooting and see if there are any helpful tips there. Troubleshooting is the most complicated part of the entire process. It will take some patience and attention to get everything finally working properly.

*note: display can be set to °F. See section 6.1 "How to change the Temp and Time settings in the program"

Chapter 5 - Using The OpenAutoclave System

The autoclave system is made to be powered from the cigarette lighter that is common in most motor vehicles. It requires 12 volts DC and draws 8 amps. For testing, it may be convenient to use a benchtop 12-volt power supply. If you opt to test with a vehicle, it is a good idea to leave the engine running during tests, or you will be unable to start the engine later!

Note: If you require the system to be powered by solar or wind power, you will need to first set up an alternative energy system that charges a 12-volt battery, and then connect the autoclave to this battery.

Also Note: If you are using your Open Autoclave without a control box or computer, skip ahead to section 4.5 (p.65) "Using the Autoclave without an electronic controller."

5.1 Run a full cycle
5.2 Download and graph the data
5.3 Using biological indicator strips
5.4 Troubleshooting
5.5 Using the autoclave without the electronic controller

Chapter 5 : Section 1 - Run a full autoclave cycle

5.1.1 Prepare the machine for a cycle

Place the insulated pot over the heater rack and press lightly into the pillow. Set the heater power switch on the control box to the "off" position.

5.1.2 Thermocouple wire's overall length must not be hot

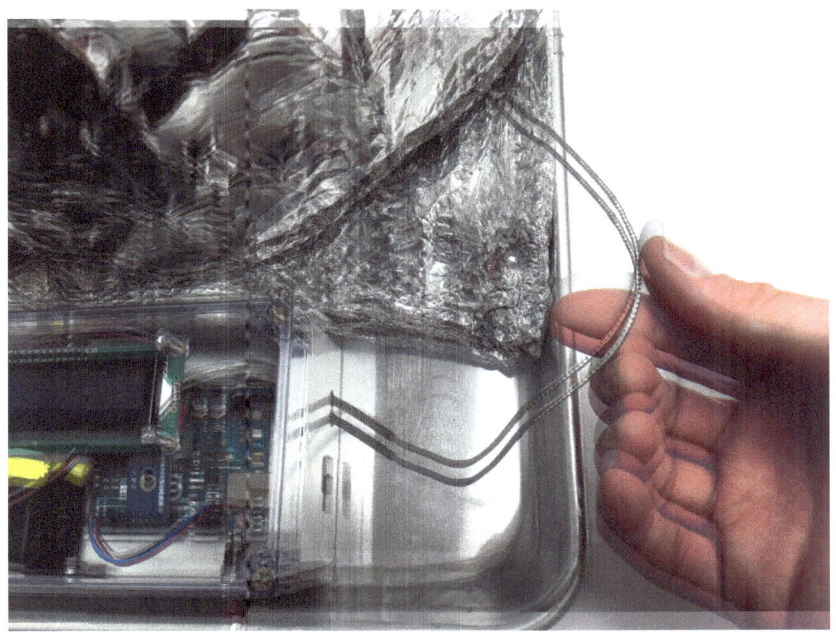

For the thermocouple sensor to take good readings, the overall length of the wire must not be allowed to get hot.

Make sure that the length of the thermocouple wire outside of the oven is free and out in the open air. It must remain cool in order for the temperature to read correctly.

Chapter 5 : Section 1 - Run a full autoclave cycle

5.1.3 Connect to 12-volt power

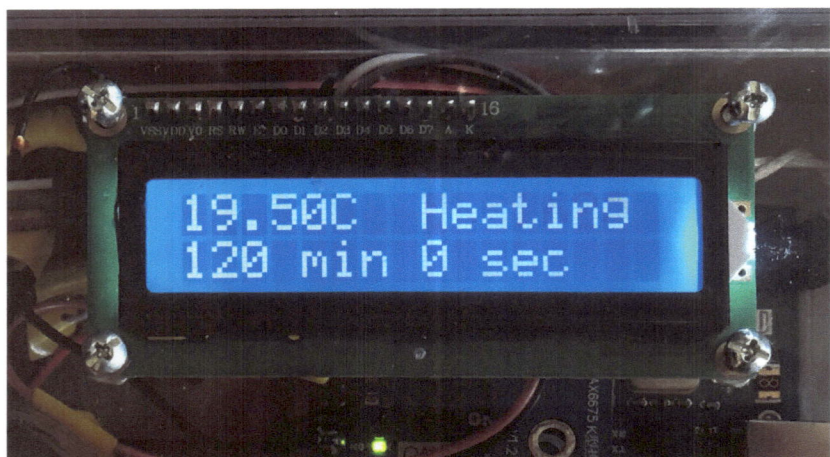

With the heater-power switch set to the "OFF" position, connect the autoclave to 12-volt power. There should be a cute "start-up" sound, and the LCD screen should display the temperature. There will also be a blinking message that says, "Heating". This message means the thermocouple is reading a temperature below the countdown-timer's start temperature.

Note: Display can be set to °F. See section 6.1 'How to change the Temp and Time settings in the program'

5.1.4 Begin the cycle

Switch the heater-power switch to "ON". Press and hold the small push button for five seconds. The screen will go blank and the computer will sound with a start-chime. The screen should then reappear to display the current temperature and a message that says "Heating." Observe the temperature for a few minutes. It will take a few minutes before the temperature begins to rise noticeably.

The downloadable Open Autoclave software is pre-programmed with the following settings:

Exposure Temperature: 160°C (320°F)

Estimated warmup time from 20°C (72°F): 50 minutes

Exposure Time at Temperature: 120 minutes

Estimated Total Cycle Time from Cold Start: 170 minutes

These settings can be changed by making small adjustments the downloaded Arduino Open Autoclave software, and then re-uploading it to the Arduino board. (See chapter 5 section 1, pg. 68) The following chart should be carefully considered when adjusting the Autoclave settings:

Temperature / Time Recommendations for Dry-Heat Sterilization

Autoclave Exposure Temperature		Estimated Warmup Time (min. from 20°C)	Recommended Exposure Time (minutes)	Total Cycle Time from Cold Start (minutes)
°C	°F			
170	338	60	60	120
160	320	50	120	170
150	302	40	150	190
140	284	30	180	210

Exposure time figures published by JADA, (Journal of American Dental Association) Vol. 122 December 1991.

Chapter 5 : Section 1 - Run a full autoclave cycle

5.1.5 Completing the cycle

Do not disconnect power to the system once the timer has begun. Disconnecting the power during a cycle will reset the countdown timer if it has begun.

The Open Autoclave is well insulated and will not cause much external heating, nor will it heat the surface on which it sits. That said, it is a small oven capable of reaching high temperatures. It should be operated under supervision, and in an area free of flammable materials or combustible chemicals.

The author conducted a series of tests of the Open Autoclave in the front seat of his car (pictured left).

While the tests were successful, the author does not recommend attempting to drive the vehicle while the autoclave is in operation! The oven top is not mechanically secured to the tray, so it slides off easily.

The author strongly recommends devising a way to secure both the base and the pot of the autoclave to a solid surface within the vehicle if you require an autoclave for use in a moving vehicle.

5.1.6 Let the autoclave cool

When the cycle alert announces the timer is complete, use caution. The contents are very HOT!

It is a good idea to let the system cool for 30 minutes or more before removing the contents, as they will be very hot.

If time is of the essence, the oven bucket can be removed, and contents left to cool in the open air for 10 minutes.

Chapter 5 : Section 2 - Download and graph the data

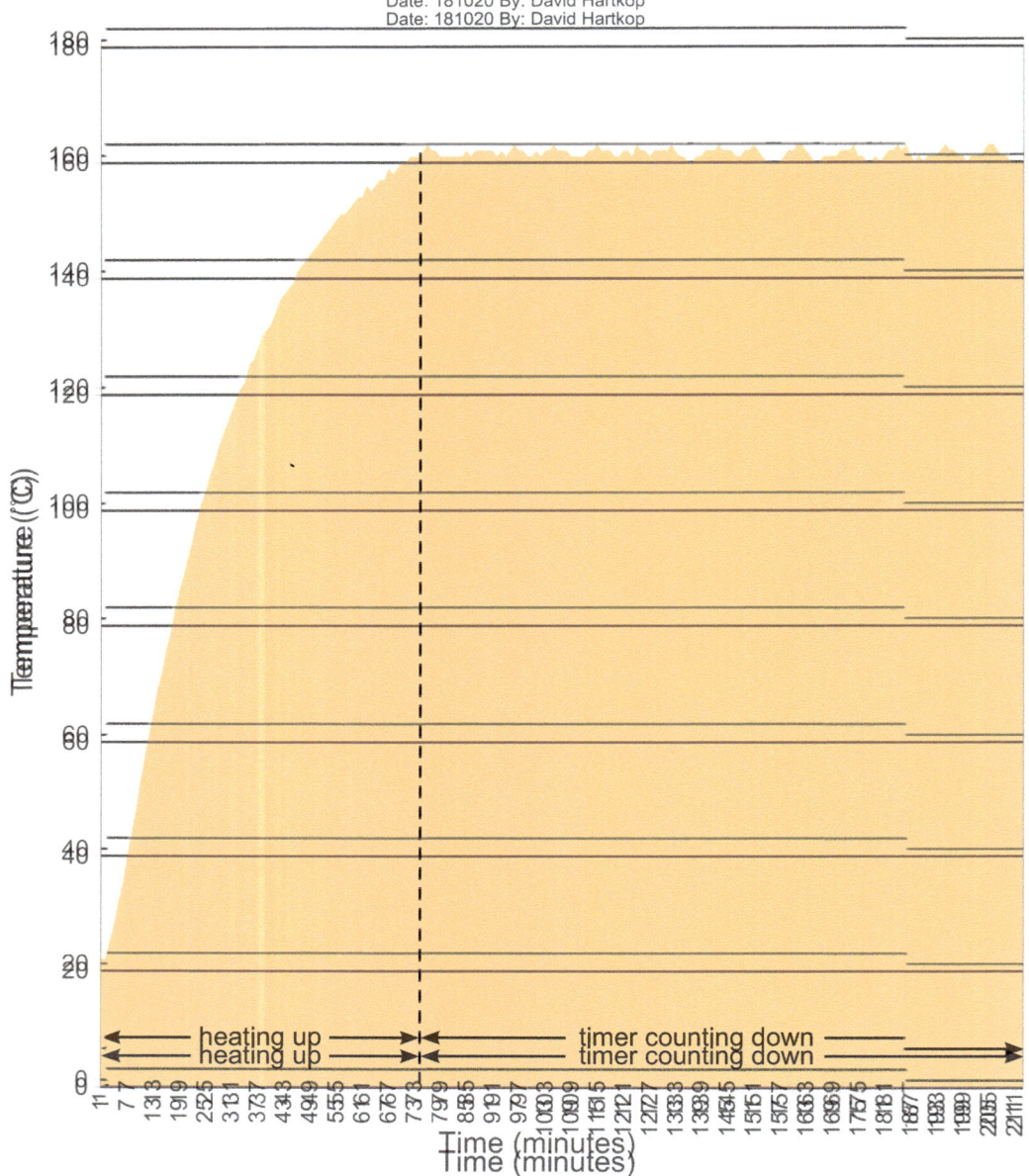

Fig. 4.2 - Temperature over time graph for a typical sterilization cycle using the Open Autoclave. Each time the autoclave runs, temperature readings are logged every 60 seconds. The data are stored in non-volatile memory on the EEPROM and can be downloaded later. The Arduino Uno can record about 8 hours of such measurements before over-writing the data with new measurements. Temperature records for each sterilization cycle may be useful or required as part of a healthcare system's quality control measures.

Chapter 5 : Section 2 - Downloaded data from the autoclave

5.2.1 Connect the Arduino to your computer

To download the data, you will need to use a computer or laptop that has the Arduino IDE software installed. This is the same software that was used to upload the software into the Arduino Uno in chapter 3:

With the 12-volt power disconnected, connect a USB printer cable to the Arduino Uno.

5.2.2 Open the serial monitor

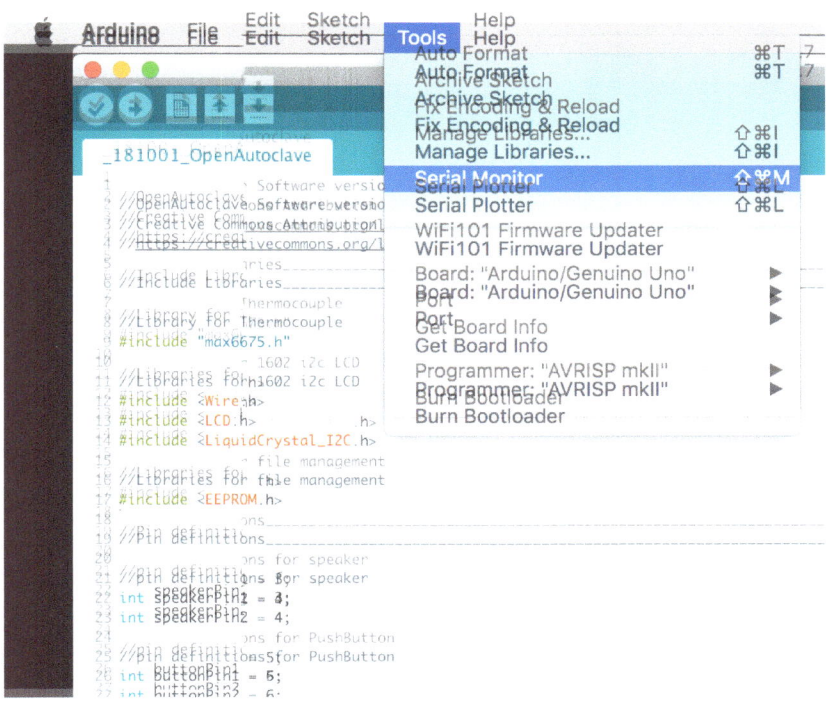

With the Arduino Uno connected to your computer via the USB printer cable, start the Arduino IDE software.

From the drop down at the top, select Extras / Serial Monitor.

Chapter 5 : Section 2 - Downloaded data from the autoclave

5.2.3 Serial monitor options

The serial monitor window provides a way to 'talk' to the Arduino computer in the autoclave.

When the serial connection is established, the autoclave will make the start sound.

A short menu of options will appear in the Serial Communications Window, shown at left.

For example, typing the letter 'i' and pressing ENTER will cause the Arduino to display a few lines of information about the current software installed.

5.2.4 Type 'd' to download all data

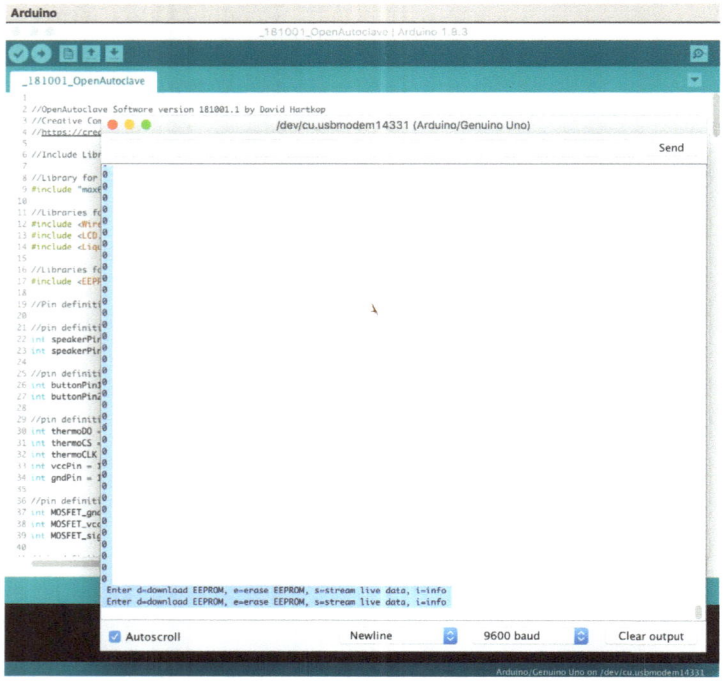

Type 'd' into the serial communication window's text field, and press "Enter" or "Return" to download all data. A large column of numbers should flow into the window and will stop after a few moments.

Select the entire column of numbers by highlighting one number and then pressing <Ctrl-a> Copy by pressing <Ctrl-c>.

Alternatively, you may right click and choose 'Select All' and right click again and choose 'Copy.'

Chapter 5 : Section 1 - Run a full autoclave cycle

5.2.5 Paste the date into a text document and SAVE it

Paste this column of numbers into a new text document or directly into a spreadsheet. Be sure to save the file to your hard drive before closing any windows.

You may wish to create a folder named 'Autoclave Data' on your desktop or in your documents directory.

I like to name my data using the day's date and a number. For instance, my first download for October 5, 2018 might be named this:

20181005_AutoclaveData01.txt

I'm obsessive, but it helps me not to get confused when I need to find a file later!

5.2.6 Deleting stored data from the Arduino Uno

The Arduino Uno has 1Kb of non-volatile EEPROM memory. This means the data remains when the power is disconnected.

The Open Autoclave program automatically logs the temperature every 60 seconds when power is connected. Each recording is placed in memory after the last, separated by the number '13'.

When the end of the EEPROM memory is reached, the EEPROM is automatically erased, and recording starts again from the first block.

You can manually erase the EEPROM. You can do this from the Serial monitor by entering the * symbol (SHIFT 8) and pressing ENTER or Return on a Macintosh. The Arduino will instantly erase its onboard EEPROM chip.

5.2.7 Interpreting the downloaded data

The column of downloaded numbers is a set of temperature readings in Celsius.

The readings were logged by the Arduino every 60 seconds during the entire time it was powered up.

Each time the Arduino is powered up, it writes one data block with the number '13' as a way to punctuate recordings. When graphed, it is easy to see breaks in the recorded data by the occurrence of the number 13 in the set.

5.2.8 Keeping a log of sterilization cycles

Keeping a log of sterilization cycles can be important. It may be required by your group or institution. In any case, a log may be kept on paper or, ideally, in a spreadsheet along with the recorded data.

I have included a sample data log spreadsheet in the folder of items downloaded from GitHub. You can maintain a log by first copying the list of temperature readings from the Arduino Serial Monitor, and then pasting the list into a column of the spreadsheet. Feel free to use this spreadsheet or to adapt it to fit your needs.

Chapter 5 : Section 3 - Using biological test strips

5.3 Use biological indicator test strips to verify your autoclave

It is common practice to test the effectiveness of sterilization using biological indicator strips. These indicators are typically a small sealed paper strip containing bacteria spores. The strip is exposed to the autoclave's heat during a normal cycle and is then mailed to a testing laboratory, where it is put through the culture process. If the autoclave was effective, no bacteria will grow from the test strip. There are several different laboratory service providers that can process biological indicators specifically made for medical sterilizers. Select one near you if possible.

Note: The Open Autoclave is a dry-heat sterilizer. This method is effective when used properly, but it requires a longer incubation time in order to verify. It may take several days to over a week to receive results from a testing laboratory.

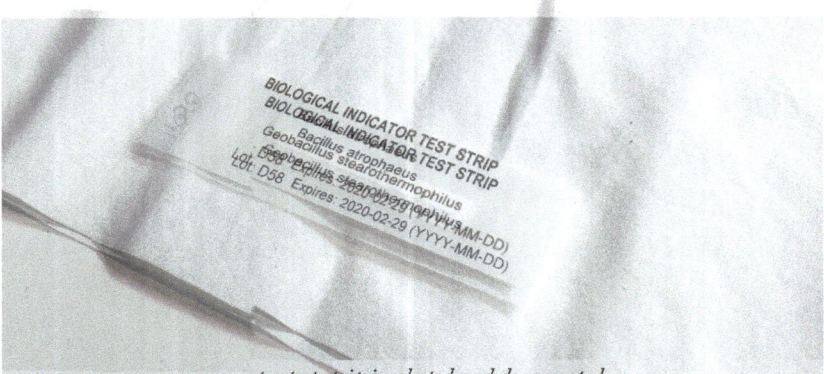

Fig. 5.3a - Biological indicator test strip is selected and documented.

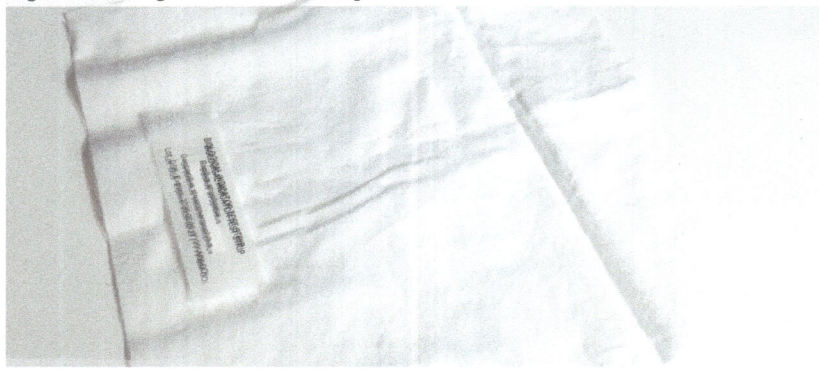

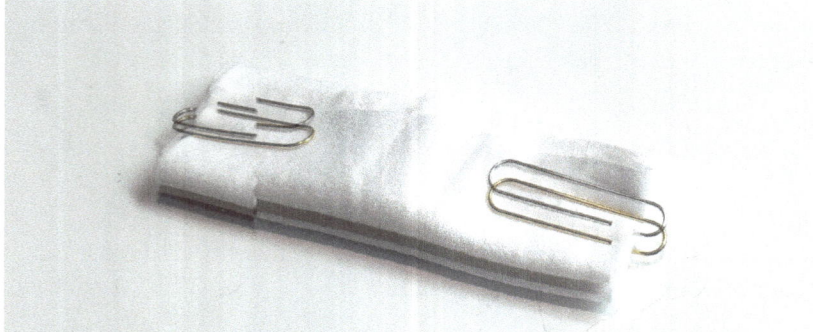

Fig. 5.3d - Biological indicator test strip shown wrapped in linen for autoclave exposure. Wrapping simulates the exposure conditions of linen-wrapped medical instruments.

Chapter 5 : Section 4 - Troubleshooting

COMPILED LIST OF ISSUES AND SOLUTIONS

I have compiled a list of problems I had when developing this system, and the solutions that were found. Hopefully this will help with your troubleshooting. Note: Unplug the power or the USB connection from your autoclave before pulling out or putting in any wires - microcontrollers are very sensitive, and hot-swapping leads to burnt out parts!

Arduino software won't start or run on my computer

If the Arduino software has difficulty starting, you may need to install a new driver on your system. This is a common issue with Windows 10. Perform a quick Google search for "Windows 10 driver for Arduino" and find some current instructions, as fixes for this are subject to change.

I can't get the Open Autoclave software to upload to the Arduino

In the Arduino program, from the drop-down menu at the top, select Tools / Port. From the flyout menu, choose one that has ".usbmodem" in the name. One may even say "Arduino UNO" in the name - if so, choose that one. If there are multiple, just guess until you get the correct one.

Arduino is plugged in to USB but does not turn on

Make sure heater switch is "OFF". USB can't power those heater coils!

LCD Lights but no picture

Check the order of the four wires. I think you have the two data wires flipped. See schematic in section 5.5 on pg. 72.

Arduino is on, but the LCD does not light up at all

Check the wires to see that they are connected between the proper pins. See schematic in section 5.5 on pg. 72.

Temperature reading seems extremely wrong

It is probably a loose wire, or the amplifier is plugged into the wrong pin(s). Try removing the thermocouple amplifier from the Arduino pins and plugging it back in. Check to see that you have it aligned to the correct set of pins. See schematic in section 5.5 on pg. 72.

Temperature reading is still very wrong

The thermocouple probe's wires may be reversed where they plug into the thermocouple amplifier. Check to see that blue connects to (-) and red connects to (+).

Temperature reading is STILL not correct

If the entire wire of the thermocouple probe gets hot, the thermocouple will read low. Check to make sure that MOST of the thermocouple wire is safely outside the heated oven bucket, and that it is out in the free air, not tucked in around the hot lower edge.

Chapter 5 : Section 4 - Troubleshooting

One more thing that throws off the thermocouple reading

If the mounting screws holding the MOSFET module come into electrical contact with the tray, it will cause the temperature to read incorrectly.

The heater coils are not heating

Are you SURE? Be careful checking because they become extremely hot to the touch. Remove the oven bucket and hold your hand over the heater rack. Check to see the heater switch is "On" and that its wires are connected. Check to see that the wires are connected to the automotive relay properly; see schematic in section 5.5 on pg. 72. Make sure you are using the "N.C." pin, not the "N.O." pin on the relay. (N.C. means Normally Closed; it is connected to the Common pin unless the coil is energized.)

I can't get the program to download into the Arduino from my computer

Make sure you have "Arduino Uno" set as the board type under the drop-down menu Tools / Board.

Nothing works, and I smell burning / see a tiny puff of white smoke

Disconnect the power. Sorry to say, but this happens sometimes - something was connected to the power backwards. If you're lucky, it was just the MOSFET module. If you're unlucky, it was the Arduino Uno. A common mistake is to connect your power supply backwards; check to make sure the contact for the TIP of the cigarette lighter is (+) and the barrel is (-).

System heats but the countdown never starts

it is possible that the coils are wired incorrectly, or that they are slightly different than those I used, and so the system is not reaching the required temperature for the countdown to start. With the power disconnected use a volt-ohm meter to measure across the two pins of the 12-volt power plug. With the heater switch set to "on" the resistance should read right around 1.55 ohms. If it is higher than 1.65 or less than 1, then something is wrong with one or more of the coils. Most likely, the heater cartridges are fine, they are just wired incorrectly. Check the schematic for wiring the coils on page 72.

Chapter 5 : Section 5 - Using the autoclave without the electronic controller

No Computer Operation

The autoclave can still be used even without the electronic controller. This requires paying attention to temperature and time. I recommend using a clock and not a stop watch and using a pen and a pad of paper as a log book. The following page of this book is an example log-book page, which may be copied and used for your manual log keeping.

5.5.1 Connect power to the autoclave and start a cycle when ready by holding down the push button for 5 seconds until it makes a start sound. Note the time and temperature in the log book. Here is an example format to use for all your autoclave log entries:

```
Date:_____
Cycle of the day #:_____
Temp Goal of _____ reached at time:_____
Completion time exposure:_____
Heater Off at:_____
```

5.5.2 Note the date and the cycle number of the day at the start of each entry.

5.5.3 When the autoclave reaches the goal temperature or above (example 160°C), note the time. This is the start time for the autoclave exposure.

5.5.4 Figure out the "Completion time for exposure:" and write the time the cycle will be complete. (Example from above: 9:37 AM start time + 120 minutes exposure time = 11:37 AM completion time.)

5.5.5 When the completed exposure time has been reached, turn off the heater switch. Write this time in the log, saying "Heater Off at: _____"

Whenever possible, it is recommended to regularly test your autoclave's sterilization performance using biological indicator test strips. See section 5.3.

OpenAutoclave Manual Operator's Log Book

Date:_____ Cycle of the day #:_____ Temp Goal of _____ reached at time:_____ Completion time for exposure:_____ Heater Off at:_____	Date:_____ Cycle of the day #:_____ Temp Goal of _____ reached at time:_____ Completion time for exposure:_____ Heater Off at:_____
Date:_____ Cycle of the day #:_____ Temp Goal of _____ reached at time:_____ Completion time for exposure:_____ Heater Off at:_____	Date:_____ Cycle of the day #:_____ Temp Goal of _____ reached at time:_____ Completion time for exposure:_____ Heater Off at:_____
Date:_____ Cycle of the day #:_____ Temp Goal of _____ reached at time:_____ Completion time for exposure:_____ Heater Off at:_____	Date:_____ Cycle of the day #:_____ Temp Goal of _____ reached at time:_____ Completion time for exposure:_____ Heater Off at:_____
Date:_____ Cycle of the day #:_____ Temp Goal of _____ reached at time:_____ Completion time for exposure:_____ Heater Off at:_____	Date:_____ Cycle of the day #:_____ Temp Goal of _____ reached at time:_____ Completion time for exposure:_____ Heater Off at:_____
Date:_____ Cycle of the day #:_____ Temp Goal of _____ reached at time:_____ Completion time for exposure:_____ Heater Off at:_____	Date:_____ Cycle of the day #:_____ Temp Goal of _____ reached at time:_____ Completion time for exposure:_____ Heater Off at:_____
Date:_____ Cycle of the day #:_____ Temp Goal of _____ reached at time:_____ Completion time for exposure:_____ Heater Off at:_____	Date:_____ Cycle of the day #:_____ Temp Goal of _____ reached at time:_____ Completion time for exposure:_____ Heater Off at:_____

Temperature / Time Recommendations for Dry-Heat Sterilization

Autoclave Exposure Temperature		Estimated Warmup Time (min. from 20°C)	Recommended Exposure Time (minutes)	Total Cycle Time from Cold Start (minutes)
°C	°F			
170	338	60	60	120
160	320	50	120	170
150	302	40	150	190
140	284	30	180	210

Exposure time figures published by JADA; (Journal of American Dental Association) Vol. 122 December 1991.

www.ideapropulsionsystems.com/OpenAutoclave
https://github.com/IdeaPropulsionSystems/OpenAutoclave

Chapter 6 - Resources

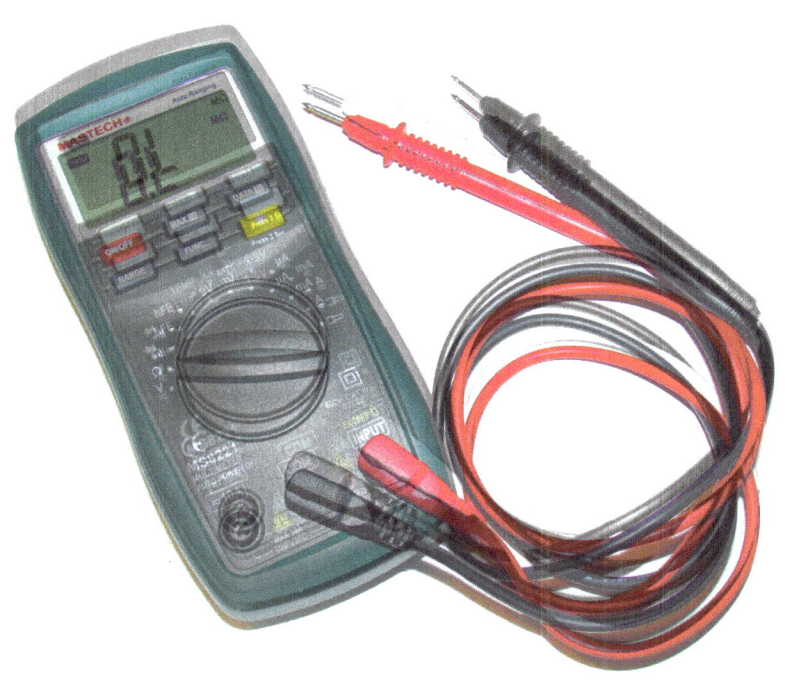

5.1 Change the temp and time settings
5.2 Volt-ohm meter
5.3 Anti-static mat
5.4 Pin connector crimper
5.5 Full electrical schematic
5.6 Diagram of autoclave system
5.7 Template for drilling electrical box lid
5.8 Template for drilling electrical box
5.9 Placement of components in electrical box
5.10 Ideas for improvements
5.11 Powering Open Autoclave with Solar

Chapter 6 : Section 1 - Change the temp and time settings

The temperature and time settings for the Open Autoclave are easy to change in the downloaded Arduino Software. Be careful to follow the autoclave exposure guidelines outlined in the chart at the bottom of this page labeled, "Temperature / Time Recommendations for Dry-Heat Sterilization." With the Arduino software opened, look at lines 7, 8, and 9 highlighted below:

Line 7 is the variable "SetPoint_heater." This is the maximum temperature the autoclave will be allowed to reach.

Line 8 is the variable "exposureTime." This is the time in minutes for which the temperature will be held. The heater shuts down after this time has elapsed.

Line 9 is the variable "displayUnits." Set this to 0 to display Celsius, set it to 1 to display Fahrenheit.

Once you have made the changes to the code in the Arduino program, save by choosing File / Save. Now, follow the steps outlined in Chapter 3 to upload the modified software to the micro-controller.

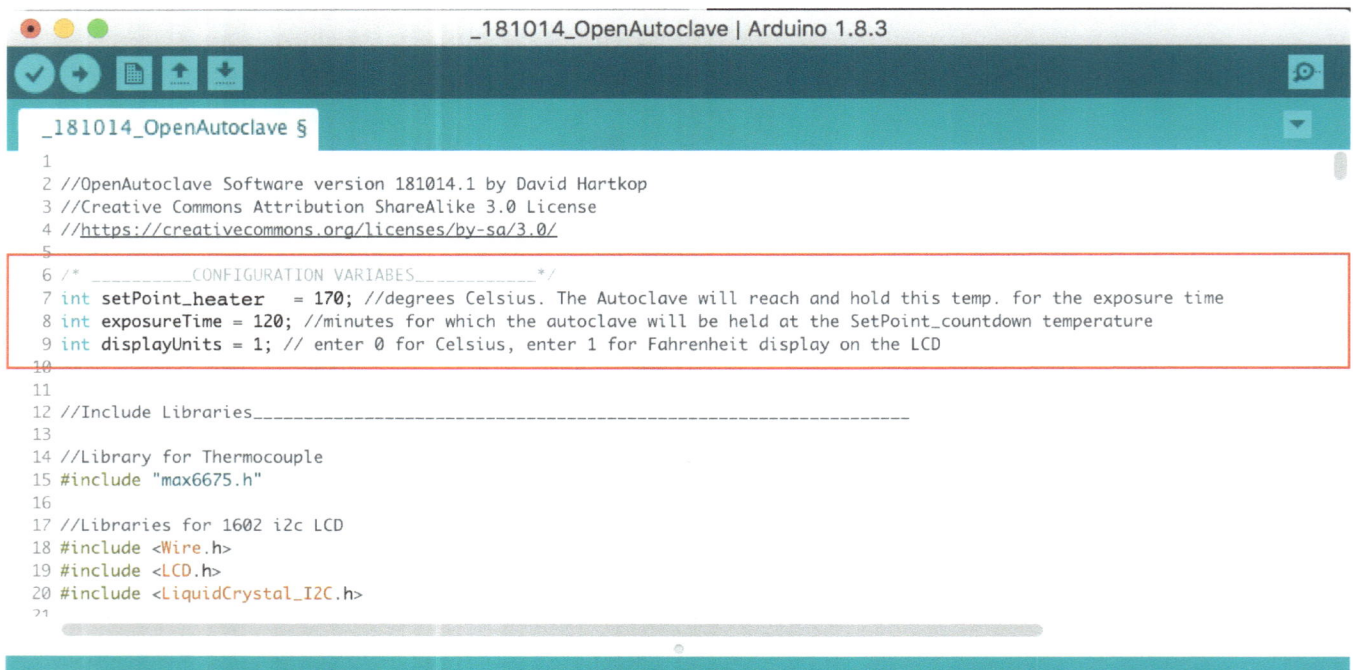

Temperature / Time Recommendations for Dry-Heat Sterilization

Autoclave Exposure Temperature		Estimated Warmup Time (min. from 20°C)	Recommended Exposure Time (minutes)	Total Cycle Time from Cold Start (minutes)
°C	°F			
170	338	60	60	120
160	320	50	120	170
150	302	40	150	190
140	284	30	180	210

Default Setting → 160 / 320

Exposure time figures published by JADA, (Journal of American Dental Association) Vol. 122 December 1991.

Chapter 6 : Section 2 - Volt-ohm meter

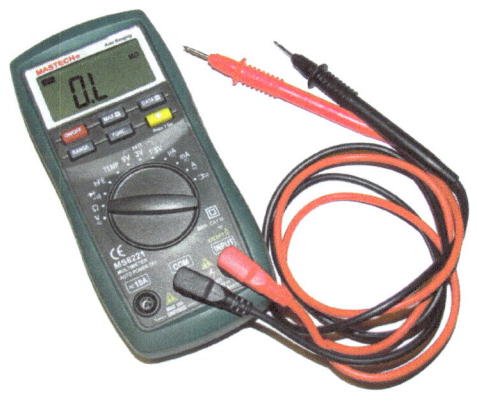

A volt-ohm meter is a handy tool when working with electronics. It lets you measure several different electrical properties, including voltage, resistance, current, and continuity.

Continuity simply means that electricity is free to flow from one point to another. Checking for continuity is essentially checking to see if two points are electrically connected to one another.

Turn the meter's main selector knob to the symbol shown in photo at left. Now, touch the probes together. The meter should issue an audible tone or 'beep' sound. The screen may also change to display a --- symbol.

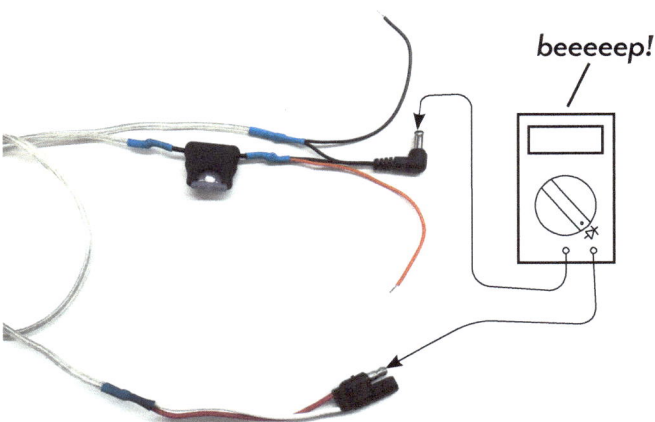

That's really all there is to it! Experiment by touching the probes to various conductive objects. This will come in handy when trying to determine what is connected to what.

The diagram at left demonstrates how to use a volt-ohm meter to check continuity between two ends of a complex cable. In this example, the meter is indicating that the center pin of a barrel connector is connected to the exposed pin of an automotive connector. **Good to know!**

Chapter 6 : Section 3 - Anti-static mat

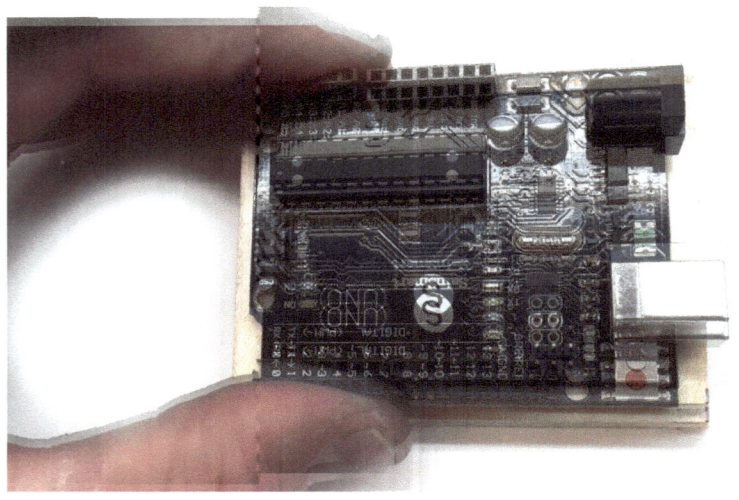

Many electronic circuits are sensitive to static electricity. If you've ever shuffled your feet across a carpet and touched a metal object, you may have been met with a small blue spark. You may not realize it, but that tiny spark that jumps from your finger to a door knob represents a discharge of hundreds to thousands of volts! This is more than enough to destroy the microscopic transistors that populate an integrated circuit.

A common way for hobbyists and engineers to prevent static-related burnouts is to use a desktop anti-static mat. Just roll out the mat on your work space, connect its jumper wire to ground, and do your work in safety on its surface.

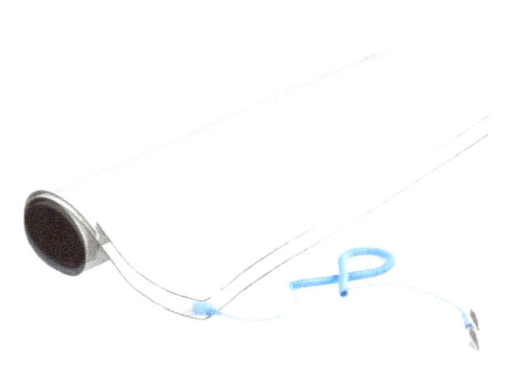

The term ground refers to a conductor that is in contact with the earth's electrical field. Typical grounds common in the home include a water pipe and the screw that holds the cover onto an electrical outlet. Many anti-static mats come with a tiny loop that is meant to be put under this screw.

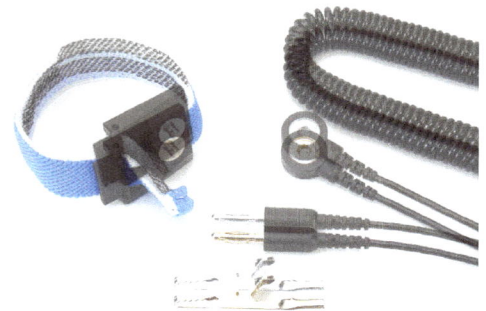

If you live in a very dry (highly static-electric) environment, it may be wise to ground your body as well.

Many anti-static mats also come with a handy wrist strap. Put it on, snap the wire to the mat's grounded receptacle, and you are ready to build!

Chapter 6 : Section 4 - Pin connector crimper

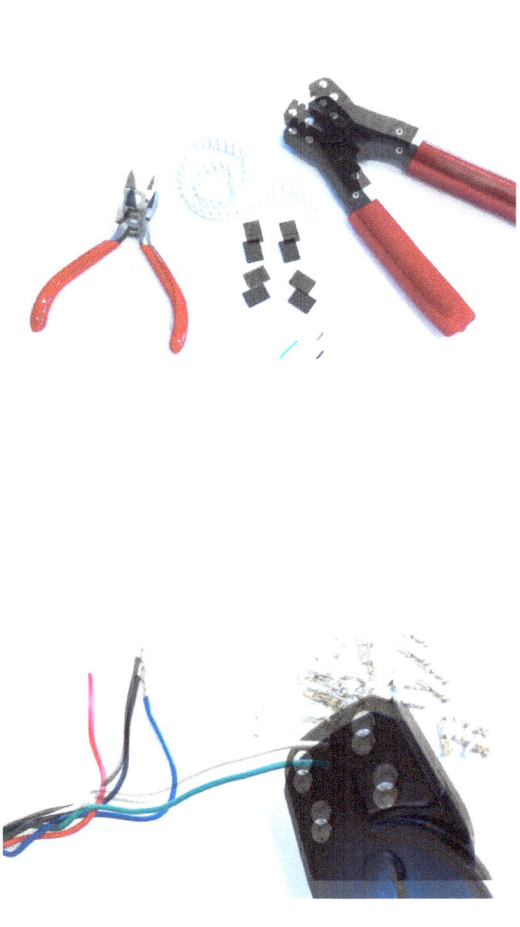

Many electronics projects involve connecting items directly to so-called header pins that protrude from a circuit board. The most common way to do this is to use a kit that includes female crimp pins and wire jumper pin header connector housings.

Crimp pins are available in rolls and require a special crimper tool. Get the tool. The housings can be purchased with many different numbers of sockets. If you are purchasing online, search for "Dupont wire jumper pin header connector housing" and for "Dupont female crimp pins".

Making up custom connectors is actually very fun and easy to do. First strip the ends of a set of wires so that about 1/8 inch is exposed. Then use the crimper tool to crimp the female pin connector onto it. The connector should firmly wrap onto the wire end around the insulation.

The final step is to "click" each wire into a socket in the connector housing. The pins can be removed later if you need by lifting the tiny retainer tabs with the end of a sharp pin micro screwdriver.

This style of connector is indispensable when working with stepper motors, sensors, switches, or other devices that need to connect to a micro-controller or other circuit board.

Chapter 6 : Section 5 - Full electrical schematic

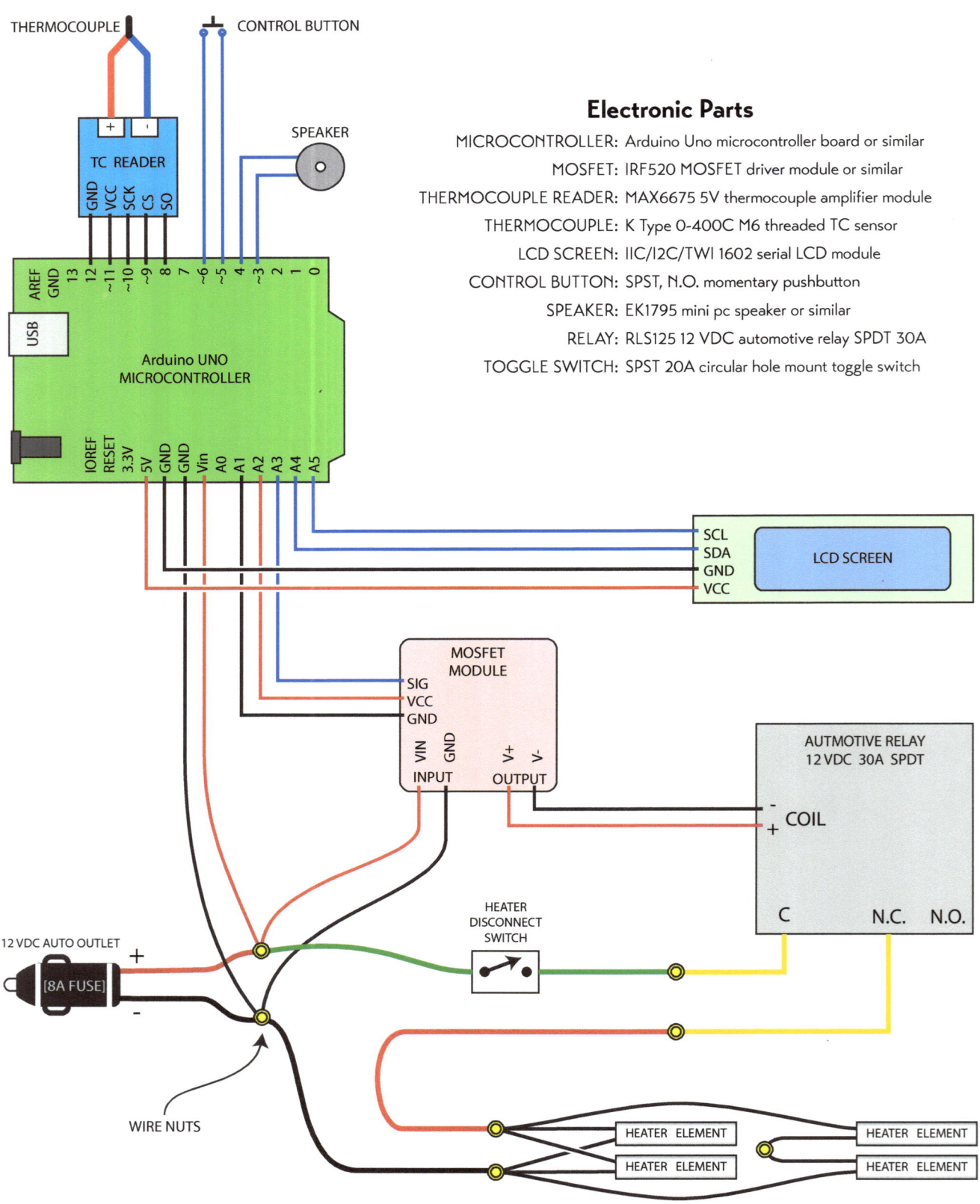

Chapter 6 : Section 6 - Diagram of autoclave system

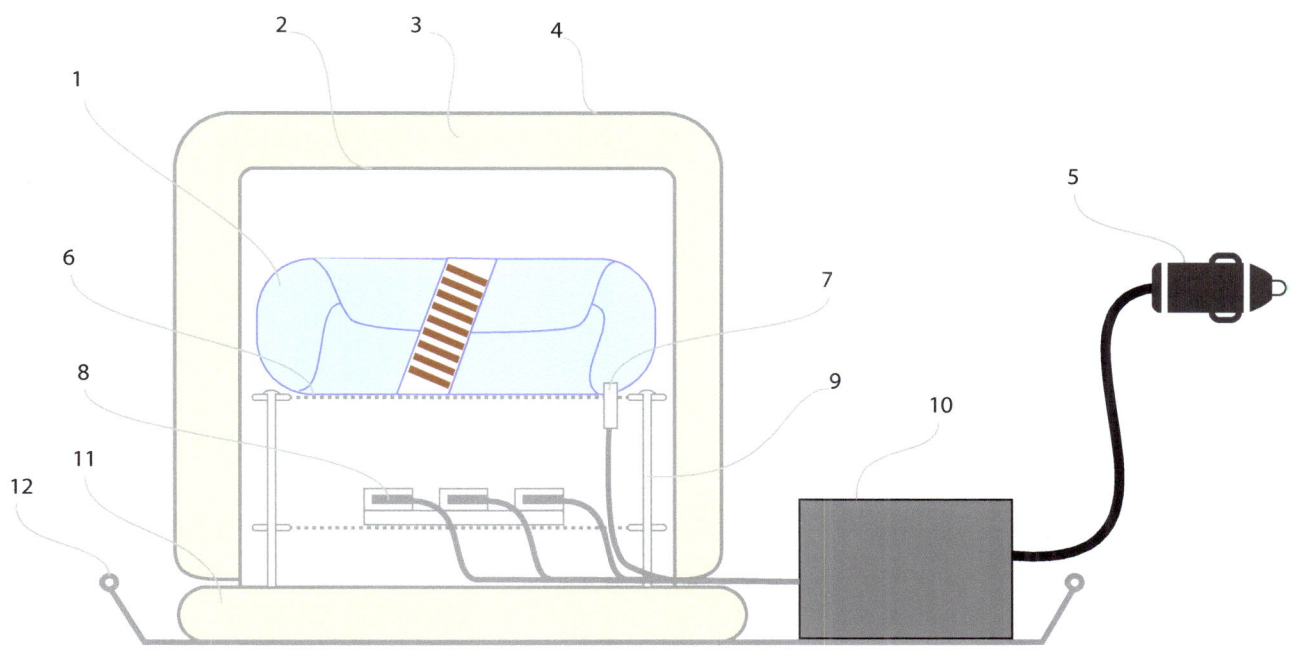

1. Wrapped medical instruments
2. Inner stainless stock pot
3. Ceramic fiber insulation
4. Heavy aluminum foil and tape outer wrap
5. 12 VDC power input connector
6. Aluminum pizza screen inner stages
7. Stainless thermocouple temperature probe
8. 40 Watt heater cartridge
9. Stainless support hardware
10. Microcontroller
11. Ceramic fiber bottom pillow
12. Full size commercial cookie sheet, aluminum

Open Autoclave: Build an open-source off-grid medical diastrument sterilizer by David Hartkop CC-BY license

Chapter 6 : Section 7 - Template for Drilling Electrical Box Lid

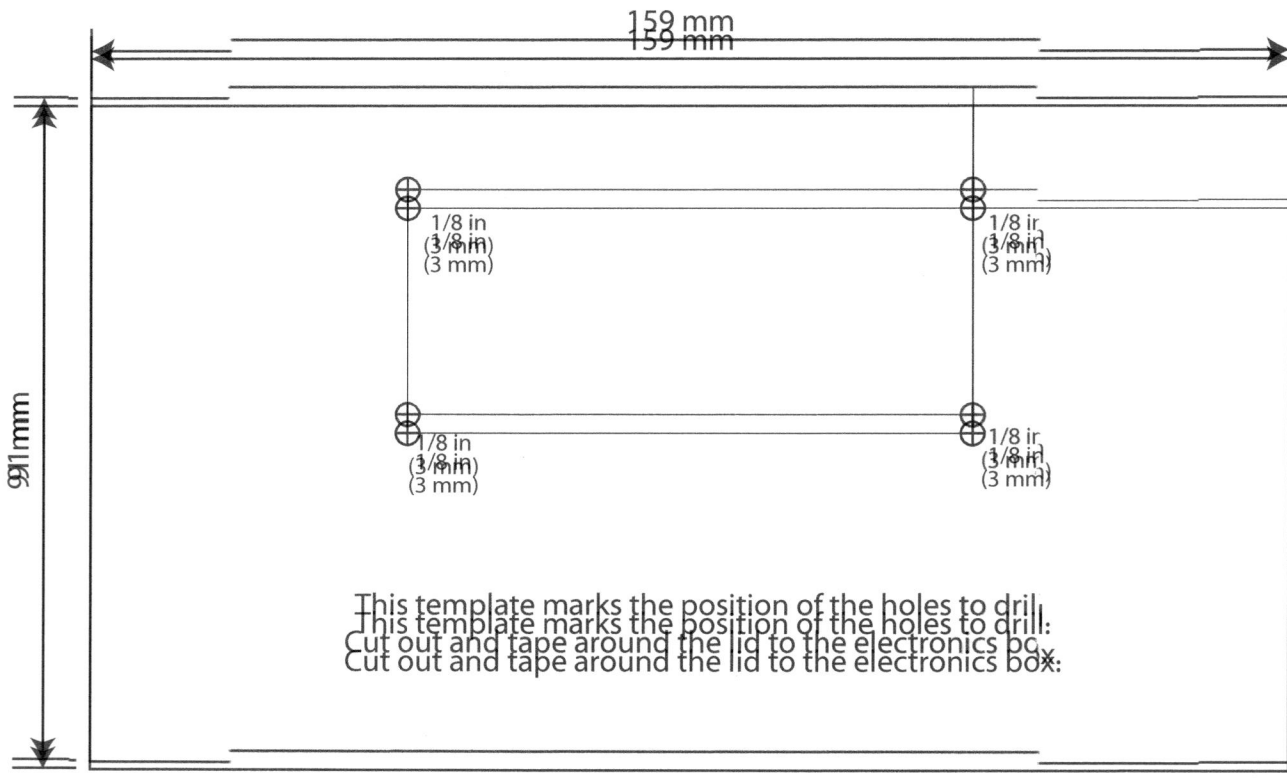

Chapter 6 : Section 8 - Template for Drilling Electrical Box

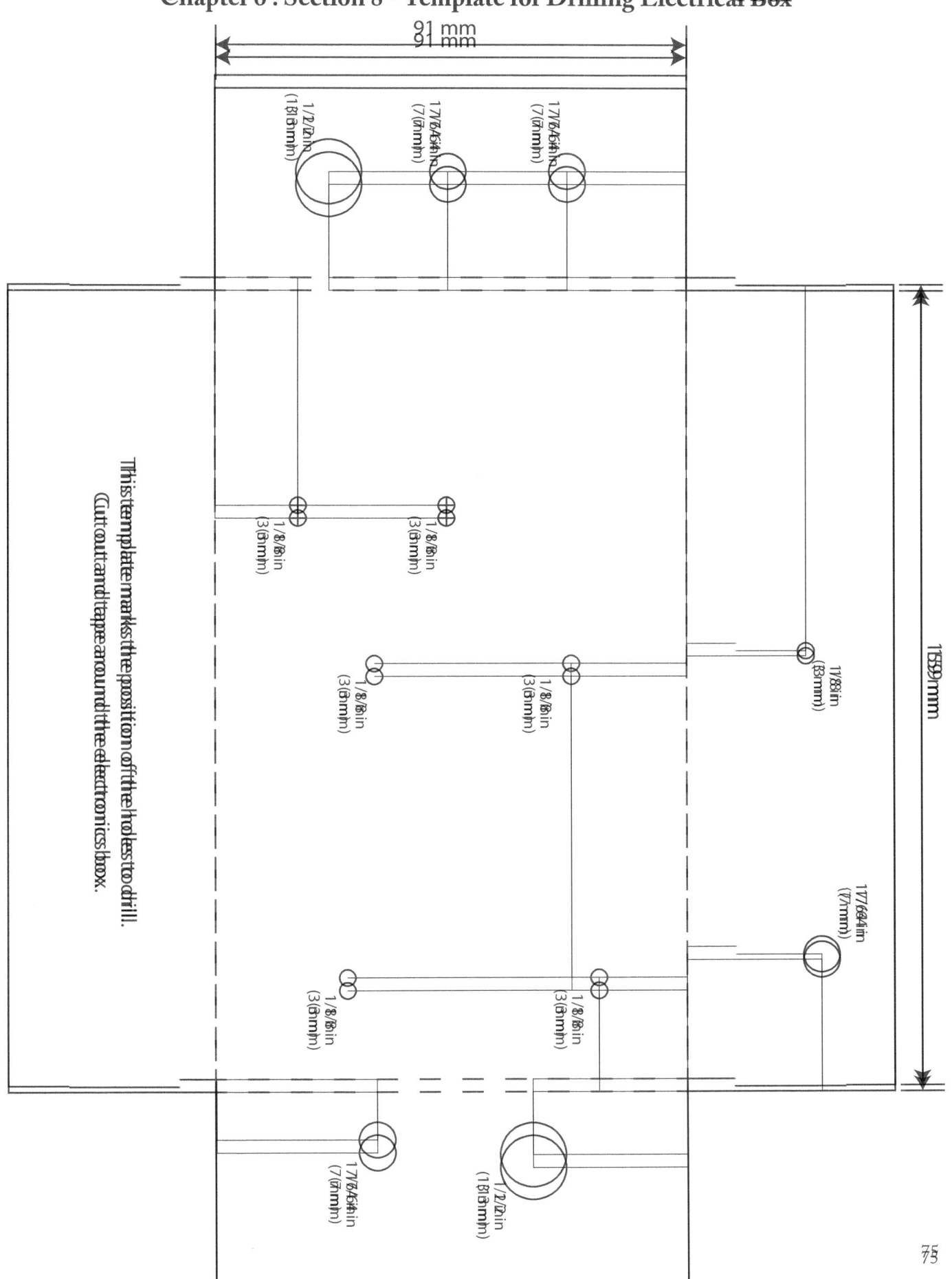

Chapter 6 : Section 9 - Placement of Components in Electrical Box

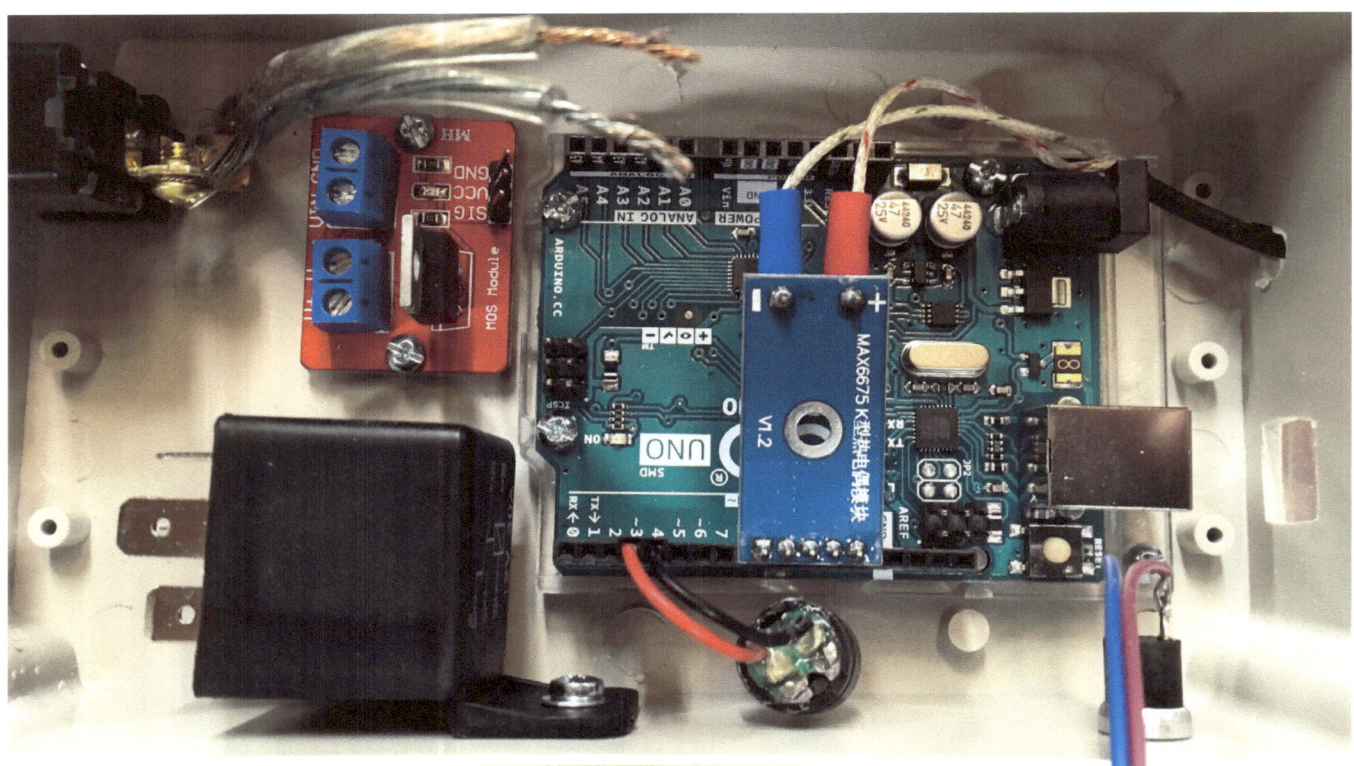

Fig. 6.9a – Piture of the main components physically installed inside of the electronics box

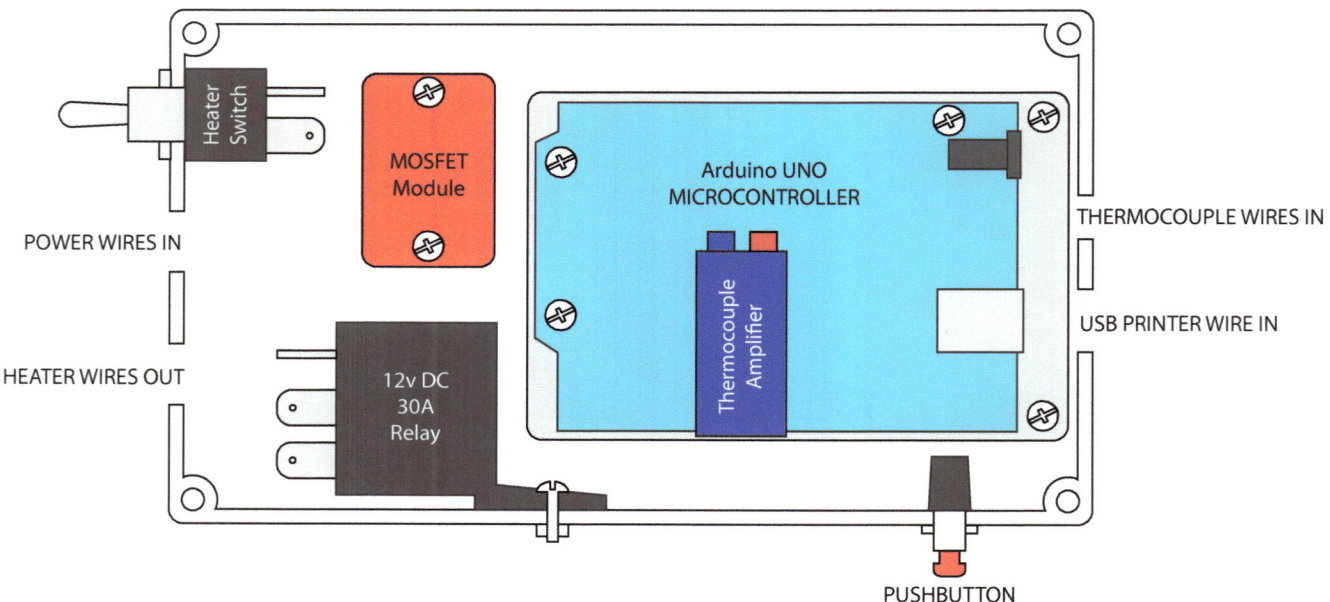

Fig. 6.9b – Illustration showing the placement of the main components inside the electrical box

Chapter 6 : Section 10 - Ideas for Improvements

- Redesign the oven's construction so it tolerates bouncing of a moving vehicle.
- Tighten up the temperature control precision with better control algorithms.
- Install a real-time clock module to log real date and time.
- Log the data to a removable flash chip rather than using the internal EEPROM.
- Use a different microctontroller that implements Circuit Python, making it easier to edit code in the field.
- Maybe there should be a better user interface?
- Create a custom PCB with thermocouple and relay already on it to speed up assembly of systems.
- Custom 3D printed electronic enclosures so that holes don't have to be drilled.
- Pre-packaged kits of the project could be made and sold.
- Set up a wireless connection to a phone app for data logging and system control.
- Set up a wiki to allow crowdsourced translation of the book into different languages.

Chapter 6 : Section 11 - Example Solar Power Setup

Figure 5.11a illustrates the best way to power the OpenAutoclave with solar energy. The solar charge controller is the key piece of equipment to purchase. A charge controller converts the extremely variable power from the solar panel into a steady voltage to charge a battery. The box also monitors the current being used by the autoclave, which lets it make decisions about how 'hard' to charge the battery.

There are many different chage controllers on the market. Some have display screens and status lights, while others are small resin-filled boxes. My advice is to choose one with decent online ratings, and expect to spend at least $50 USD on it.

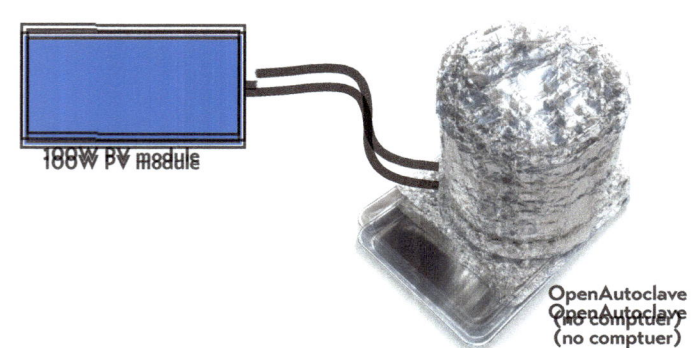

Figure 5.11a - Solar power operation using charge controller

Figure 5.11b illustrates a stripped-down method for powering the OpenAutoclave with solar energy. A 100 watt photovoltaic module is wired directly to a set of three of the 40 watt heater cartridges. The optimal wiring of the heater cartridges for this mode of use is shown in figure 5.11c below.

We found experimentally that the power from a PV panel alone is too variable to directly power an Arduino thermostat system. The use of buck-converters or more sophistocated regulated DC-DC converters was found to be problematic. The systems would work, but unreliably. The only converters that work well cost more than a car battery and a charge controller.

If you want to use solar without a battery or charge controller, use the manual method described in Chapter 4.5 (pg 65) covering "No Computer Operation."

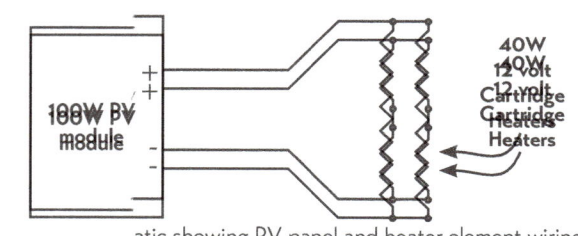

Figure 5.11b - Solar power operation with no charge controller and no computer.

Figure 5.11c - Electrical schematic showing PV panel and heater element wiring.

Acknowledgements

Many thanks to my wife Lindsay for her encouragement, patience, and layout skills. Also to Laura Horton and Michele Hartkop for proofing and arrangement suggestions.

About the Author

David Hartkop is a graduate of the school of Film and Television of Loyola Marymount University, class of 2000. He has worked in commercials, feature film, and eventually found a home in the world of digital special effects. Other projects include solar coffee roasters, 3D printing with metal clay, and experiments with flying spherical cameras on drones. Hobbies include sci-fi reading, designing contraptions, running, and storytelling with dramatic flair. David lives with his wife and son in Medford, Oregon.

Share Your Experience

If you or a team have completed and are testing the Open Autoclave, please complete my survey page. Your independent work is invaluable to improving and ultimately deploying the Open Autoclave where it is needed most.

The survey page also includes tips and suggestions about how you can share your project and improvements with the maker community.

www.IdeaPropulsionSystems.com/OpenAutoclave/YourBuildSurvey

Links to Online Resources

Idea Propulsion Systems

www.ideapropulsionsystems.com/OpenAutoclave

GitHub

https://github.com/IdeaPropulsionSystems/OpenAutoclave

Google Drive

Shortened URL: goo.gl/39sAaL

https://drive.google.com/open?id=11Lix_WFGX7QbaYG3f3S69cg09V7xdMFZ

Dropbox

Shortened URL: https://tinyurl.com/yay7c2de

https://www.dropbox.com/sh/n99b3ym8ilzcc91/AAD0lpurYqdqwlUdi6cMt5jMa?dl=0Dropbox

www.ingramcontent.com/pod-product-compliance
Lightning Source LLC
Chambersburg PA
CBHW051915210526
45473CB00006B/2025